The End of Fossil Energy and Per Capita Oil

By John G. Howe

ISBN 978-0-9962054-5-0

Cover photo credit: NASA Goddard Space Flight Center Image by Reto Stöckli (land surface, shallow water, clouds). Enhancements by Robert Simmon (ocean color, compositing, 3D globes, animation). Data and technical support: MODIS Land Group; MODIS Science Data Support Team; MODIS Atmosphere Group; MODIS Ocean Group Additional data: USGS EROS Data Center (topography); USGS Terrestrial Remote Sensing Flagstaff Field Center (Antarctica); Defense Meteorological Satellite Program (city lights).

Dedication

To my children and all the children of the world …
they will need all the resources we can conserve.

Howe Engineering Company
Fossil Energy and Depletion
Issues and Solutions

February 1, 2016

Re: COVER LETTER for 5th EDITION BOOK

Dear Reader,

You have in your hands my comprehensive self-published book. The book is the fifth edition and my final summary of a 12-year study to understand and explain the complex interactions between energy, population, environment, and economics.

My conclusions lead to grave concerns regarding the continuation of our high-energy lifestyle. This book includes possible solutions for personal preparation, how we can downsize energy needs, and still provide the basic necessities for ourselves.

In a nutshell, the quantitative facts regarding our predicament focus directly on the second-half of the two-lifetime oil age, beginning now. Most of today's global crises are in some way related to keeping Americans moving in their petroleum-fueled life style. As the dominant source of world wealth, which is largely borrowed from the future, Americans have supported and extended a plateau of world oil consumption since the concept of "Peak Oil" that first became popular in 2005. In that period alone, the world consumed another 300 billion barrels of oil. In fact, in the lifetime of a 25-year old today about one-half trillion barrels of oil (or about half of the total world oil so far in the oil age) have been consumed, also contributing to greenhouse gases and climate change.

Now, the "days of grace" are over. The world's remaining oil endowment is rapidly depleting at a rate of one billion barrels every eleven days. We must prepare for drastic changes in the next ten to twenty years. Unfortunately, this geologic fact is disguised behind a temporary glut and low price primarily because the average American motorist is not responding directly to overproduction from suppliers who are desperate to maintain their income.

The most significant thing you can do, other than personal preparation, is to educate yourself and spread the word to others, preferably VIPs, politicians, and media. My book is listed on Amazon for $15.00, but whenever possible I will give free copies to anyone interested in promoting this urgent message. Time has run out!

John G. Howe, President

Howe Engineering Company • howe@megalink.net
www.solarcarandtractor.com • YouTube.com/Howe Triple Crisis

Foreword by Colin Campbell

Thank you for sending me the 5th Edition, forwarded by Fintan O'Connell. I think you are entirely right. The prime chapter of the Oil Age spans little more that two lifespans as you illustrate on the cover. The First Half saw the rapid and excessive expansion of everything, including human population, fuelled by easy and cheap energy. But the Second Half which dawns will see a corresponding contraction. The transition through which we are living is a time of great tension as indeed already witnessed with many demonstrations, riots and revolutions around the world. People become resentful of soaring food prices and falling employment. They blame their governments, not realising that the situation is imposed by Nature. The turmoil in the Middle East, North Africa is interesting because they are rich oil countries, but that delivered massive immigration and an excessively wealthy and powerful elite. Their populations far exceed the carrying capacity of the barren lands, and the pressures mount. The countries are artificial constructions with most frontiers drawn by Britain and France in the First World War but now there is a reversion to tribalism, with some of the revolutionaries exploiting religious differences. The State of Israel is a further tension probably because of its close ties with the financial world. Nigeria too is interesting as Arabs from the deserts of Libya now march south as part of the Boko Haram movement, evidently casting their eyes on the green forests. They probably don't realise that the country uses its oil revenue to import about one-third of its food. Its conventional oil peaked in 2005 and its deepwater oil will do so around 2017, so food imports will soon dry up and the population shrivel.

I think the solution is a reversion to regionalism as people again have to learn to live on whatever their area can support, growing their own food, which implies a radical fall in population. There are signs of this in Europe. The Scots are close to leaving the so-called United Kingdom, and there are similar moves elsewhere especially in Spain and Italy. The European Union is under pressure to break up, and there are growing moves to restrict immigration. I expect there will be a similar response in the United States. It is a financial empire built indeed largely on the global expansion of trade and industry during the First Half of the Oil Age. I am sure that the people of Maine could look forward to happy and successful lives once they break their ties with Wall Street and turn to your solar tractor.

I was born in 1931, one year after the First Lifespan illustrated on the excellent cover of your paper opened, and am spared the Second, which will be much more difficult. My grandson who lives in Norway may find himself having to catch fish from an open boat in mid winter as did earlier generations there. Its oil production peaked in 2001 at 3.2 Mb/d but has since fallen to 1.5 Mb/d and is set to continue to decline at about 4% a year. But it is a large country blessed by a relatively small population of 5 million with a density of 16 per km^2, albeit living close to the Polar Circle.

The Oil Age is in fact a brief span of time in an historical context. It is indeed a fascinating subject to study and observe. Congratulations on your splendid work.

Acknowledgments

This book is essentially a solitary project. There is no editorial staff. Much of the content and methodology are new and reflect my personal views and therefore, except where directly cited in the text, footnotes and end notes are not used. There is not time or help available for indexing. Because my book is free wherever possible, the usual retailer-distribution-publisher-author income flow is completely circumvented. The frequent use of bold text is to help highlight especially pertinent thoughts and/or facts.

Most fortunately my daughter, Virginia (Ginny), is an extremely competent self-employed publisher with skills in all aspects of visual media. She has helped self-publish several academic books in my sporting goods research and development career as well as the first three editions of "The End of Fossil Energy." Her websites are: VirginiaHowe.com and McIntirePublishing.com.

As we move into our second printing I would also like to acknowledge my friend, author, and historian Don Zillman. His astute copy editing has helped immensely to clean up my engineer's spelling and punctuation.

My only other direct acknowledgment is, of course, Debbie, my wonderful wife now for forty years. She has been a patient, long-suffering, 24/7, sounding-board for every aspect of my evolving thinking about energy, Peak Oil, and the precarious prospects for future civilization. Our two boys, now in their thirty's, were born long before we became concerned about the future for an ever-growing population in a finite world.

Instead of direct reference, the thinking and conclusions in my book reflect a broad spectrum of influence by many respected authors and activists. I will start by acknowledging the titles cited in the bibliography and/or directly in the text. Sometimes I agree, sometimes I don't, but these are my mentors.

Secondly, with the new World Wide Web and social media communication there is an avalanche of web and blog sites. Some are directly related to published works. Others reflect national and international non-governmental-organizations (NGO's) with substantial staffs and an underlying need to maintain a donor or subscriber base without offending their supporters. Many of the best sites are by committed individuals who are as alarmed as I and attempt to communicate, educate, and establish a dialogue with like-minded cohorts. Unfortunately, this exchange of ideas is hopelessly overwhelmed. Each personal thread has many followers. Some have time to contribute to open blog sites. In my opinion, these are "choir rehearsals" that do not have a chance of reaching beyond their immediate audiences which have the web addresses, time, and inclination to partake in the activity.

A list of internet sites, groups, or individuals with direct influence and bearing on my thinking follows in no particular order:

ASPOIreland.wordpress.com, Dr. Colin Campbell
Peakoil.com, Dan C.
Peak-oil.org, Association for the study of Peak Oil and Gas (ASPO-USA)
Thegreatchange.com, Albert Bates
Peakprosperity, Chris Martinson
Survivepeakoil.blogspot, Peter Goodchild
Pewlicanweb.org, Gutteriez
Firstfinancialinsights.blogspot, Terry McNeil
Endofenergy.blogspot, Terry McNeil
Peakoilbarrel.com, Ron Patterson
Theenergyexchange.com
Artberman.com
Oildecline.com, Alex Kuhlman
Cluborlov.com, Dmetry Orlov
Oilcrisis.com, Ron Swenson
Petroleumtruthreport.blogspot, Art Berman
Peakoilblues.org, Kathy McMahon
TheHillsGroup.org
SKIL.org, Jack Alpert
Resilience.org, Post Carbon Institute/Richard Heinberg
Postcarbon.com, Post Carbon Institute

Kuntsler.com, James Howard Kuntsler
Forksoverknives.com, Personal-energy facts. From the other Dr. Colin Campbell
Drkinesa.blogspot.com
Euanmearns.com
Crudeoilpeak.info, Australia, Mushalik
Communitysolution.org, Arthur Morgan Institute
Energy.utexas.edu, Carry King
Growthbusters.com, Dave Gardner
Energyskeptic.com, Alice Friedemann
Ourfiniteworld.com, Gail Tverberg
Laetusinpraesens.org, Anthony Judge
Resourceinsights.blogspot.com, Kurt Cobb (*Prelude*)
Transitionnetwork.org, Rob Hopkins
Ourworld.unu.edu, United Nations University
EIA.gov, U.S. Energy Information Association
IEA.org, International Energy Agency
WorldEnergyOutlook.org, IEA publications
PopulationMedia.org, Bill Ryerson (see YouTube/Howe Triple Crisis)
CompassionateSpirit.com, Keith Akers
OKEnergytoday.com, Bob Waldrop
WorldPopulationBalance.org, David Paxson
LivingEconomiesForum.org, David Korten
CounterCurrents.org

Preface

The end of 2015 marked the tenth anniversary of that pivotal time in history when conventional world oil extraction leveled off at seventy-five million barrels per day (Mb/d). Since 2005, as simple economic theory would have predicted, the price of oil increased five-fold thus supporting ever-more expensive extraction efforts to keep the world supplied with its most critical energy source. **Appendix A shows oil as absolutely fundamental to the support of our modern industrial life style including the acquisition and utilization of all other energy sources.** It is the only energy source which provided and can continue to support a food supply barely adequate for seven billion humans. Oil is the only form of stored fuel which can provide the modern transportation we take for granted. Modern warfare works only with copious oil. **Also, the steady growth of an oil-based economy is the backbone of a buy-now, pay-later, debt-based financial system.**

Yet we, supposedly intelligent humans, cannot accept the reality that **the bulk of the petroleum-based epoch is absolutely limited to not longer than two, eighty-year lifetimes,** the first of which is now behind us. Meanwhile we waste the precious remaining few years which are critical for a transition to a long-term sustainable future. As the saying goes: "Are humans on the earth smarter than yeast in a petri dish?" To further confuse matters, the increase in the price of oil made possible a temporary remission as we accessed non-conventional sources with ever-more expensive technology. **At ninety dollars per barrel and three trillion dollars a year, the extraction and sale of oil is just as important as a source of income for dwindling numbers of world suppliers as it is for a growing population of consumers. There are fewer and fewer customers who can afford more expensive oil thus disrupting the precarious balance between supply and demand.** As a result of the temporary glut the price dropped dramatically beginning in 2014. Still, for a growing populace around the world, being priced out of the market is manifested as a hungry belly instead of just a decline of lifestyle and mobility.

This is my fifth attempt to update this growing human predicament, find legitimate answers, and explain the story to anyone who will listen. I do this because I feel I have a comprehensive and accurate picture to present, plus glimpses of the few possible ways left to postpone or transcend our fate. **In this information age available through the World Wide Web, Google, and Wikipedia, there is an overwhelming morass of facts that, taken together quantitatively, are extremely disturbing. Why isn't the public getting the message?** This fifth-edition book will wade into the details while there still may be time to affect the outcome.

To further hide the energy-crisis, the overarching enormity of the imminent end of the oil age is shoved aside and over shadowed by a number of competing subjects. All are more or less directly related, but subservient to the ubiquitous dominance of modern civilization's precarious dependence on oil. In no particular order:

The other two fossil fuels, **coal and gas**, still dependent on oil for acquisition.

The seemingly **intentional obfuscation** of definitions for oil, condensate, gas, etc.

The energy and **economic cost to access** the remaining oil.

The future and fuel supply of **nuclear energy**, but only as a source of electricity.

The segue to a limited **solar-electric future** based on weak, sporadic sunlight.

The **need for oil in every part** of a high-tech complex society.

The relentless **growth of world-wide population** all needing to be fed.

The longer time frame of **climate change**, a direct result of burning fossil fuels.

The daily human impetus to **maximize income** and maintain business as usual.

The economist's mantra that a **critical shortage** will lead to a substitute.

The hope that the decline of oil will be resolved by **scientific breakthroughs.**

The relentless **drumbeat of world geopolitics and war**, largely because of oil.

The steady increase in debt at all levels because **oil-based growth has ceased.**

The daily stream of **misinformation by a confused and/or biased media.**

The ingrained nature of politicians and **leaders to promise a rosy future.**

The **human tendency to avoid and ignore dire predictions.** Instead, live for the day.

It's this last observation which is most disturbing because it forestalls any possible chance for a desperate eleventh-hour course-correction while there still may be time. This is especially true for "experts," think-tanks, and governmental agencies who focus on several, but not all, of the above competing subjects and offer naive, inadequate, or short term solutions. This gives the listener an easy out and eliminates the desperate need for more individual clamor, personal involvement, or group action on a macro-basis.

If we focus the conversation directly on oil, and **especially the enormity of U.S gasoline consumption, instead of blaming wild market swings on the Chinese economy,** all the other subjects fall into place. The pieces of the puzzle fit together and the ominous, emerging picture becomes very clear.

Prefaces to Previous Editions

As an attempt at continuity, I will repeat the first several paragraphs of the prefaces from my four earlier editions all with the same title, The End of Fossil Energy, but with increasingly disturbing subtitles.

PREFACE TO THE FIRST EDITION (2004)
A Plan for Sustainability

I'm sitting in a rural New England farmhouse in February 2003 trying to start this project for the fifth time in the last 10 years. The setting could not be more ideal in my comfortable, senior years as a retired engineer, farmer, part-time historian, and sometimes skeptic. Yet, something seems terribly wrong. It's a cold winter day, and in the last few minutes the back-up oil burner has switched on to smooth out the heat from our woodstoves. Now, silence is disturbed again by the school bus delivering the neighbor's children. A huge logging truck just trundled down the road with wood destined for who knows where—it varies, pulp for one of the paper mills, saw logs for Japan, or maybe just biomass for chipping and electrical cogeneration. A skidder is still roaring into the late afternoon, and a jet has left the east coast swinging west overhead. Fortunately, the snowplow is only sanding today, and our driveway is clear thanks to a gallon of gas in our 50-year old John Deere tractor.

The sound of energy intrudes, as does the TV reporting the latest on the pending war in Iraq (which just happens to have over 100 billion barrels of conventional oil reserves, the second largest in the world after Saudi Arabia). The war news is frequently interrupted by ads for things we must have, like 4 x 4 pick up trucks and snowmobiles.

Our president mentioned hydrogen fuel in his State of the Union address the other night. It's hard for me, as an engineer, farmer, and manufacturer of bicycle-powered generators, to imagine a hydrogen-powered school bus, skidder, snowplow, tractor, jet plane, etc. I am sure he means well but wonder if he understands the language of BTU's, watts, kilowatt hours, and calories as well as the details of non-fossil energy sources like solar, nuclear, wind, geothermal, biofuels, and hydro.

PREFACE TO THE SECOND EDITION (2005)
The Last Chance for Sustainability

It is now February 2005, two years after starting the first edition of *The End of Fossil Energy and a Plan for Sustainability.* Since then the world has consumed another 55 billion barrels of oil, including 15 billion in the U.S., which is twice the 7 billion expected to be in ANWR. The Market price has doubled to almost 60 dollars per barrel. Oil company profits are at record highs but far fewer oil fields are being discovered each year. There are previously discovered fields coming into production and about one-half of the world's original endowment is still in the ground, but worldwide demand has clearly outpaced production.

China and India have experienced unexpectedly high economic expansion and have become major players in the world energy markets. Due to continued violence, Iraq is only producing about 1.9 million barrels of oil per day, well below its pre-war level of 2.5 million.

PREFACE TO THE THIRD EDITION (2006)
The Last Chance for Survival

In less than one year since our second printing in 2005, it is apparent that the subject of fossil-energy has moved into the mainstream of public awareness. The following items highlight the transition of energy topics from a fringe group of gloom and doom alarmists to accepted legitimacy.

The term peak oil has become the common phrase for media and public attention. In the 12 months from January 1, 2005 to January 1, 2006 crude oil jumped 36 % from $45.00 per barrel to $61.00. By January 15, 2006 it had increased another 9% to $67.50. In the same period, natural gas increased a whopping 100% from the $6.00 range to over $12.00 per thousand cubic feet. By January 25, 2006 it had settled back to $8.25 due in large part to the extraordinarily mild winter in the U.S.

On December 8, 2005 the House Energy and Commerce Subcommittee on Energy and Air-Quality held the first full-scale congressional hearing on peak oil. This bipartisan caucus is co-chaired by Roscoe Bartlett (R-Maryland) and Tom Udall (D-New Mexico). Resolution 507 was co-sponsored by 16 prominent congressmen and begins with the first paragraph:

Expressing the sense of the House of Representatives that the United States, in collaboration with other international allies, should establish an energy project with the magnitude, creativity, and sense of urgency that was incorporated in the "Man on the Moon" project to address the inevitable challenges of Peak Oil. (See **www.energycommerce.house.gov** for a complete transcript.)

PREFACE TO THE FOURTH EDITION (2014)
The End of Fossil Energy, What Next? It's Up to You!

The 4th edition (2014) was self-published only in a spiral-bound manuscript format. This was a summary of my continuing work and presentations since 2006. The attempt was to enlist reader participation, local printing, and networked exponential exposure of the 118 page comprehensive effort. As would be expected, this dream did not go very far. Much of this 4th edition is refined in Parts II, III, and IV, and is combined with my more recent thoughts regarding per capita oil and gasoline consumption in Part I.

Contents

Definitions of Frequently Used Terms

This is a brief list of abbreviations and acronyms used in this book:

b/p/y	barrels per person per year
Bb	Billion barrels
Bb/y	billion barrels per year
cpf	child per female
cpm	child per male
L/A	lead acid
Mb/d	million barrels per day

PART I

FOCUS on PER CAPITA OIL

The end of the short, high-energy oil age looms before us in so many ways. Yet we avoid the subject and fail to understand how we Americans are central to this perilous subject.

Instead, we dwell on much longer-term climate change while we motor down the road frantically consuming the liquid finite fossil fuel which will be critical to our and our children's survival. (See Appendix A)

The thoughts presented in this book are not political or vague conjecture. They are based on hard unequivocal math, physics, and evidence.

We start our study by first turning our focus back towards ourselves and our grossly inordinate oil consumption. If there's any chance of shaping our destiny, we must first identify and tackle the root problem.

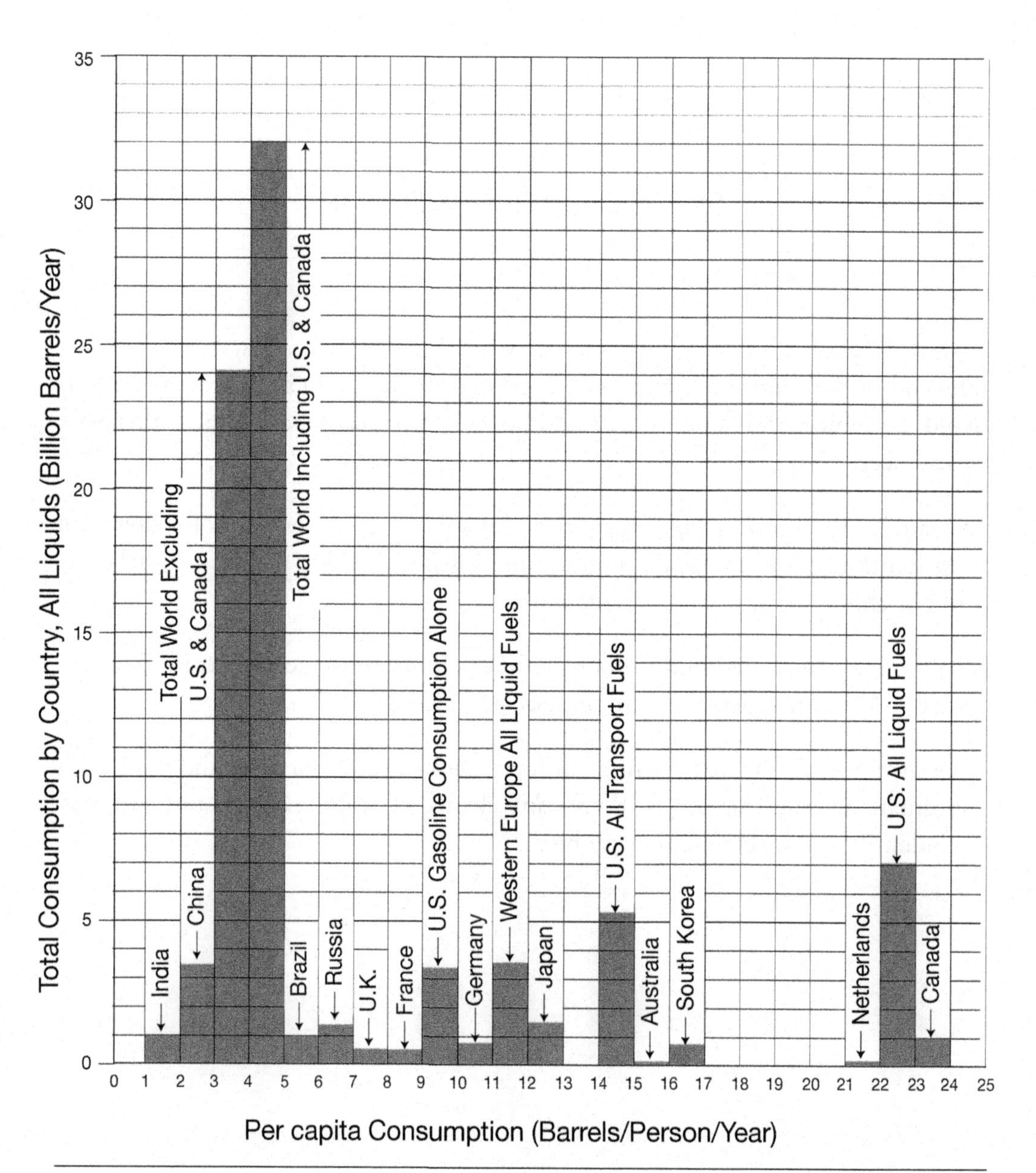

FIGURE 1 World Oil Consumption by Country and Per Capita Consumption

CHAPTER 1

World Liquid Fuel Consumption Highlighting U.S. Gasoline

On the facing page, Figure 1 is a bar-chart showing total and per capita oil consumption for the top twenty oil-consuming **countries** in the world **with the U.S. by far the most dominant. Also included in this bar chart are the total consumption and per capita rates for just U.S. gasoline, and another bar for all U.S. liquid fuels used specifically for transport including diesel and jet fuel.** The numbers used in Figure 1 are duplicated in Table 1 in rounded numbers. Accuracy may vary slightly depending on date and source. Several other countries and Western Europe as a bloc are included for reference: (For reference use google or wikipedia.)

Also included in Figure 1 are:

- Total world consumption of 60 countries **(including the highly-skewed contribution from the U.S. and Canada)**, all consuming 32 billion barrels per year (Bb/y) of all petroleum liquids with a population of 7.2 billion. The average world per capita consumption is therefore 4.4 barrels per person per year (b/p/y).
- Total world consumption and per capita consumption **excluding the U.S.** The numbers are then reduced to a total world consumption of 25 Bb/y used by a population of 6.9 billion people with a per capita consumption of 3.62 b/p/y.

Most revealing is the bar in Figure 1 which shows 3.3 billion barrels of oil per year (Bb/y) used in the U.S. by 310 million Americans ... just for gasoline! This would be equivalent to a stand-alone country of 310 million people and a per capita total oil consumption of 10.7 barrels per person per year (b/p/y). In more familiar units, an average of about 400 gallons of gasoline are used in the U.S. each year by every man, woman, and child. Does each American really need one and one-quarter gallons of gas to make it through the day? It could be argued that a small fraction of U.S. gasoline and diesel is exported to Europe and Asia, but that in no way changes the argument and absolves the American motoring bloc from being the third most dominant factor in the world petroleum market. Surely we need to better understand the specifics and implications of this egregious anomaly.

TABLE 1 Numbers used in Figure 1

Country	Consumption (Mb/d)	Consumption (Bb/y)	Population (billions)	Per Capita Consumption (b/p/y)
United States total	19.2	7.0	0.31	22.6
U.S. transport only	14.5	5.3	0.31	17.1
China	10	3.6	1.3	2.8
U.S. gasoline only	9	3.3	0.31	10.6
Japan	4.4	1.6	0.13	12.4
India	3.3	1.2	1.26	0.9
Russia	3.2	1.2	0.15	7.9
Brasil	2.6	0.95	0.2	4.75
Saudi Arabia	2.8	1.0	0.03	34.5
South Korea	2.3	0.84	0.05	17.8
Mexico	2.2	0.8	0.12	6.7
Canada	2.4	0.92	0.04	23.0
Singapore	1.4	0.51	0.005	102.0
Indonesia	1.3	0.47	0.25	1.9
Australia	1.0	0.36	0.023	15.6
Iran	1.7	0.62	0.08	8.0
Germany	2.4	0.86	0.08	10.7
France	1.8	0.65	0.07	9.3
UK	1.7	0.43	0.06	7.2
Italy	1.5	0.54	0.06	9.0
Spain	1.4	0.51	0.046	11.0
Netherlands	1.0	0.36	0.016	22.5
Total West. Europe	9.8	3.6	0.33	10.9

Notes:

- Saudi Arabia and Singapore are more egregious cases of per capita consumption than the U.S., but are not included in Figure 1 because their populations and total oil consumption are much lower.
- **U.S. gasoline consumption alone (3.3 Bb/y) is nearly equivalent to the total of all liquid fuel consumption for the six dominant countries in Western Europe. (3.6 Bb/y).**
- U.S. liquid fuel consumption (7 Bb/y) is equal to 65% of the **total** liquid fuel used by all the dominant countries in Asia (10.9 Bb/y); China, Japan, Russia, India, South Korea Iran, Saudi Arabia, Singapore, and Australia.
- U.S. gasoline consumption alone (3.3 Bb/y) is far more than the total liquid fuel used in the three other dominant western hemisphere countries (2.67 Bb/y); Brasil, Mexico, and Canada.

American's profligate use of gasoline represents almost one-half of total U.S. oil consumption and **most of the other half is also used for movement of people or goods,** but with jet fuel or diesel. The next table is an approximate breakdown of all U.S. petroleum consumption including critical uses for agriculture, national defense (about 1 Mb/d), and heating oil (about 0.3 Mb/d):

TABLE 2 U.S. Liquid Fuel Consumption

	Oil Used	
Fuel Type	(Mb/d)	(Bb/y)
All gasoline	9	3.3
Distillate (diesel, heating oil)	4	1.5
Jet fuel	1.5	0.5
Natural gas liquids including propane	1.0	0.4
All other uses plus exports and imports	3.6	1.3
TOTAL	19.1	7.0

Combining all forms of petroleum-based transportation (movement) of people or goods provided by gasoline, diesel, and jet fuel, reveals graphically in Figure 1 that the ubiquitous and uniquely-mobile American way of life uses about 5.3 billion barrels of oil per year. This translates to over 14 barrels (602 gallons, 1.67 gallons every day) per year for each and every American for a steadily-depleting energy resource **that can never be replaced or substituted for by any other form of energy; not natural gas, not electricity.** Forget natural gas, hydrogen, biofuels, or electric cars and trucks. Although technically possible, the required time, capital investment, infrastructure, energy input, storage difficulties, and limited range of alternative forms of travel preclude an acceptable future for a nation that is already sinking under a mountain of debt.

ECONOMIC CONSEQUENCES

A quick look at long-term (eia.gov) history shows that U.S. oil consumption has only started to decline in the last decade. This demand-destruction began concurrent with the five-fold increase in the cost of gasoline and oil **and also included a ten-percent increase in population in the same time frame.** As the price of oil increased from the twenty dollar per barrel to the hundred dollar range, the extraction of non-conventional oil sources like deep off-shore and "fracked' tight shale became profitable and justified their immense capital investments. **But the higher cost for all petroleum-related products gradually forced more and more Americans living on social welfare, social security, fixed, or minimum-wage incomes into economic distress because they no longer could afford to pay for other discretionary needs beyond their first priority of fuel and food.** The fracked natural gas resurgence helped some by reducing the cost of home heating and electricity, But low-cost natural gas does not pay for gasoline and food with forty-seven million Americans now on food stamps. Shelter comes next in priority but is squeezed by what is left for mortgage payments or rent after fuel and food expenses.

Family expense for gasoline

A simpler, personal micro-view of this complex economic interaction is to ponder how each of increasing numbers of marginally-economic Americans, working part time for minimum wages and/or collecting some form of special assistance, can afford to underwrite the soaring macro-costs to drill deeper in more inhospitable places for the earth's remaining oil. The take-home pay of an eight-dollar per hour worker is less than $15,000 per year. This steadily increasing number of economically disenfranchised consumers cannot afford gasoline costing $3.50 per gallon or $5,600 for a family of four each member using 400 gallons per year.

The last thing most Americans want to "throttle-back" is the necessity, freedom, and ingrained love affair with the automobile. But, change they have, which, combined with more efficient cars at all income levels, has somewhat reduced consumption by the most dominant customer in the world. The result is "demand destruction" from the bottom up of the income ladder which, in turn, lowers the price of liquid fuels down closer to balance the much-higher costs necessary for extraction (production) of remaining non-conventional sources. **The clear conclusion is that waning U.S. gasoline demand, the third most significant bloc in the world (after U.S. or China total oil consumption), directly influences and determines the price of world oil rather than the conventional wisdom of the other way around.**

Quantitatively, in round numbers from "google" sources, "peak driving" in the U.S. occurred in 2004/2005 at 24,000 miles per household, 13,000 miles per licensed driver, and 9,000 miles per capita. In the ten years following, all three indicators have declined about ten percent despite a population increase of over thirty million.

In the last months of 2014, Saudi Arabia finally grew tired of losing market share and income beginning back in 2005 when the American motoring public reached peak U.S. gasoline consumption. This Saudi movement has been exacerbated in the last five years by a surging supply of higher-priced, non-conventional sources like hydrofractured ("fracked") tight shale and Canadian tar sand supplies. **Why should the Saudis support higher-cost U.S. non-conventional extraction only to lose control of their national income?** The Saudi decision to over-supply the world with oil at less than $50 per barrel, down from the $100 range, instantly threw the oil markets and short term market speculators into a spin. Gasoline from $3.75 down to the $2.00 per gallon range suddenly released an extra jolt of discretionary per capita purchasing power of over $500 per year for the average American. Of course, much of this windfall is spent on a resurgence of miles traveled thus reinvigorating the American penchant for burning through precious remaining (U.S. or World) oil as quickly as possible with no concern whatsoever for the not-too-distant future.

TOTAL WEALTH, PER CAPITA WEALTH, AND MEDIAN INCOME

Another way of arguing the premise that the unique combination of **U.S. oil consumption, population size, and residual personal wealth is by far the most dominant force that "drives" the world oil markets,** is to focus on individual consumer spending power and per capita median income for all twenty major oil-consuming nations in U.S. dollars (Table 3). This data is readily available from the internet through google references:

TABLE 3 Spending Power of Oil Consuming Nations

Country	Consumption (Bb/y)	Population (billions)	Income (per capita)	Personal Wealth	Total Wealth (trillions)
United States	7.0	0.31	$15,480	$143,000	$44.3
China	3.6	1.3	$1,786	$11,000	$14.3
Japan	1.6	0.13	$10,840	$125,000	$16.2
India	1.2	1.26	$616	$6,500	$8.2
UK	0.43	0.06	$12,349	$130,000	$7.8
Germany	0.86	0.08	$14,023	$90,000	$7.2
Italy	0.54	0.06	$6,874	$121,000	$7.2
France	0.65	0.07	$12,443	$95,000	$6.6
Spain	0.51	0.046	$7,284	$93,000	$4.2
Brasil	0.95	0.2	$2,247	$20,000	$4.0
Canada	0.92	0.04	$15,181	$90,000	$3.6
Mexico	0.8	0.12	$2,900	$23,000	$2.7
South Korea	0.84	0.05	$11,350	$45,000	$2.2
Russia	1.3	0.15	$4,129	$16,500	$2.4
Australia	0.36	0.023	$15,026	$90,000	$2.1
Indonesia	0.47	0.25	$541	$8,000	$2.0
Netherlands	0.36	0.16	$14,450	$121,000	$1.9
Iran	0.62	0.078	$3,115	$16,600	$1.3
Saudi Arabia	1.0	0.03	$4,702	$22,000	$0.7
Singapore	0.51	0.005	$7,345	$113,000	$0.6

THE CRUX

By combining the above facts regarding personal wealth, income, population, and driving habits, the complex nexus of remaining world oil, consumption rate, price, and the future of our ephemeral, two-lifetime, oil-based society comes sharply into focus and can be accurately predicted.

The only country that has the combination of individual wealth (including debt), population, and per capita income to significantly dominate world oil consumption is the United States at 7 billion barrels per year (Bb/y). This conclusion is exacerbated by rapidly growing wealth disparity. Many Americans still have enough residual wealth and income to afford higher-priced gasoline. The GDP and public confidence are maintained as this wealth flows down though the economic ladder from the "haves" to the "have-nots." Rich or poor, a significant fraction of American wealth is lost (burned through) as liquid fossil fuels at every level. The remainder of the oil age will be a delicate death-dance between how much longer (and at what price) poorer Americans will still be able to purchase oil (specifically gasoline) and how desperately the remaining producer-nations (including the U.S.) are willing to fuel the American driving addiction in order to continue supporting their own oil-based economies.

China is a distant number-two to the U.S. 7 Bb/y with **total** oil consumption of 3.6 Bb/y. But its five-times larger population dictates that Chinese per capita income, including gasoline consumption is less than 3 b/p/y, and is therefore not the culprit as is the U.S. Other advanced countries like Japan, Canada, Australia, the E.U. countries, and South Korea still have significant per capita income and modern lifestyles, but their much smaller populations, short driving distances, and, in most cases, high gasoline taxes, make them minor players in the world oil scene. Other wealthy countries as in Scandinavia are also not large enough to make the top-twenty list and be major factors in the world oil market. **How much longer the oil age will last will be determined primarily by the balance between decreasing American personal wealth, motoring lifestyle, and the cost of remaining world oil.** We're talking about personal wealth as opposed to national wealth. The U.S. debt is directly approaching twenty trillion dollars, with another seventy-trillion dollars of unfunded long-term liabilities like future social security, medicare, and veteran's benefits. All were premised on the myth of everlasting economic and population growth with no consideration of diminishing resources. See Chapter 10 for further discussion of energy vs. economic growth.

Geopolitics

Most of the world is struggling to exist on an average of three barrels per person per year (b/p/y) which declines steadily as population continues to grow. The result can be seen everywhere, geopolitically, as growing anarchy. Disenfranchised young men and angry mobs revolt to hang onto some form of acceptable existence and western-world lifestyle. The remaining oil exporting countries are often ruled by monarchs or authoritarian leaders who isolate themselves behind grandeur and guards with little left for the masses except devastated countryside. **Meanwhile, much American wealth (and lives) are spent to maintain the flow of oil from the lands of 3 b/p/y marginal consumers to us, the 22 b/p/y happy motoring Americans who still control the oil markets.**

Remember, every drop (gallon, barrel, ...) of oil burned for motive power and heating is gone, forever! Nothing is left to show for this frantic consumption of the world's most vital finite resource except crumbling roads, worn-out cars, pollution, and greenhouse gases. The world production of food for seven billion people is also directly dependent on this downward spiral, but at least there's the extra step of people being fed before millions of years of sunlight energy, conveniently stored in the oil, vanish entropically into the surrounding void.

Since the concept of peak oil first became popular in 2005, another 320 billion barrels of oil have been used in the world. This is about 1/4 of the total oil used since the beginning of the oil age. This was a lost decade we could have used to begin an orderly decline of consumption and transition to an acceptable future. Instead it was wasted as many experts sought to prove that "peak oil is dead."

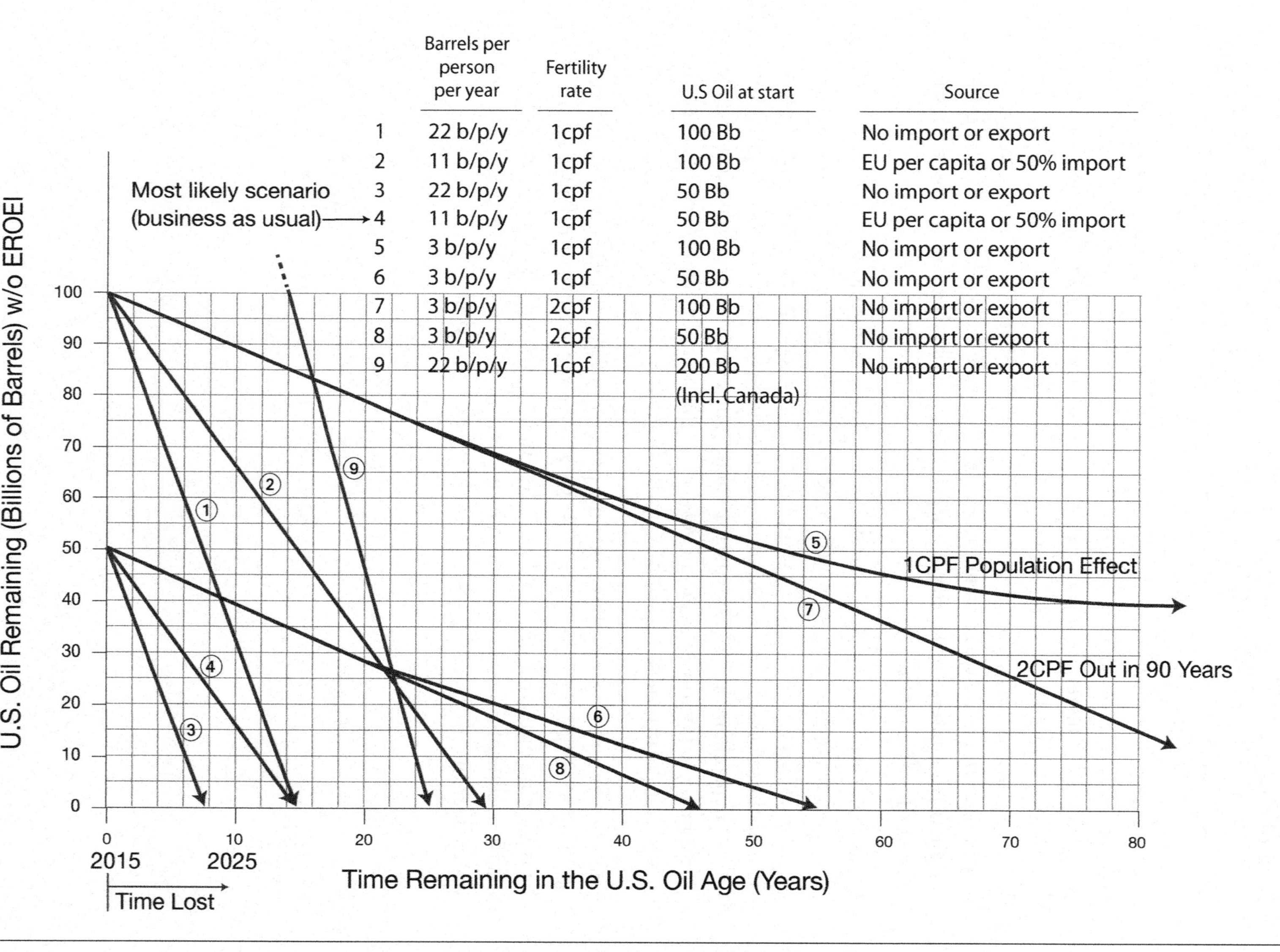

FIGURE 2 Nine Scenarios for the End of the U.S. Oil Age

CHAPTER 2

How Much Longer Do We Have?

If we focus directly on American gasoline consumption as the overwhelming elephant in the room, how much longer can the U.S. lifestyle continue? Is there any thing we can do about it? First, we need to better understand how much oil is left in the world and the U.S. There is great misinformation and hype about how much oil remains, at what quality, and the rate and cost world consumers and Americans can afford to frivolously burn it up; precious finite energy gone forever. This subject is covered in detail in Chapter 4 for all resources, but will be briefly introduced here. A quick "Google" of oil reserves in the world and by country shows world proved reserves at the end of 2012 to be **1,668 billion barrels of which only 35 or 2% is in the U.S.** This infers about fifty years of oil age remaining for the world at 32 billion barrels per year, and **five years left for the U.S. without any world oil import or export.**

The definition of "proved reserves means that even these estimates may be optimistic because:

- No consideration (discount) is included for this **steadily declining ratio of oil recovered divided by the oil required for extraction** (energy returned on energy invested or EROEI which is comprehensively defined in Wikipedia.com).
- No consideration is given to the **soaring costs to extract the remaining, increasingly-unconventional oil.** Keep in mind that the American lifestyle was founded, and functions only, on easily-obtainable, conventional, light, sweet crude.
- No consideration is given to the historic **fact that over 200 billion barrels have already been extracted from the lower 48 states and Alaska.** Half of this was prior to the Hubbert-predicted U.S. peak in 1970–72, and another 100 billion barrels in the last 45 years as extraction rates continued to decline until 2010. **Since then, the increased market price justified ever-more expensive technology which added more non-conventional oil to keep Americans rolling on a flat plateau of consumption.**

- No consideration is given to **the history of extensive domestic drilling already behind us since WW II when the U.S. almost totally supplied the world.** In the short history of the oil age there are, and have been, more wells drilled in the U.S. than the rest of the world combined. You can just squeeze so much juice out of an orange.
- No consideration is given to **alternative fuels** such as biofuels, a 1 Mb/d component of the 19 Mb/d U.S. liquid fuel supply, **and natural gas liquids** which contribute another 3 Mb/d. And, it should be emphasized that ethanol is also an equally-negative drain on other liquid fuels because its EROEI is well-established at close to unity.

NINE SCENARIOS SHOWN IN FIGURE 2

In order to avoid endless argument about how much oil remains available in the U.S. to perpetuate the gasoline-intensive American lifestyle, nine possible scenarios (combinations) are summarized in Figure 2. The nine trend-lines show, in the Y-axis, the U.S. domestic oil remaining for each scenario after a specific number of years shown in the X-axis. This is just unequivocal, grade-school arithmetic, similar to being able to predict how many miles and gallons are left for a trip or heating season at various specific rates of consumption.

Closer contemplation of each of these nine scenarios will answer many questions and respond to bogus panaceas presently offered to avoid confronting the critical energy mess we are in. **Unfortunately, all of the better paths forward require drastic changes and upmost urgency to get started. Each year of time lost after 2015 further reduces any hope of pursuing a longer way forward because the remaining indigenous U.S. oil shown in the Y axis is decreased by another 7 Bb/y as long as there is no import or export (3.5 Bb/y with fifty-percent import as now).**

The primary alternatives used in the nine scenarios in Figure 2 are:

- **U.S. per capita consumption**

  What if we were to reduce our uniquely-extravagant rate of 22 b/p/y to a level of 11 b/p/y, typical of western Europe or Japan? Better yet, what if we were to reduce U.S. consumption rate to the average world rate (without U.S. and Canada) of 3 b/p/y? Keep in mind, China is presently at 2.8 b/p/y and India averages 1 b/p/y.

- **Remaining U.S. oil at the start of each scenario**

  Rather than argue about whether or not we have 35 billion barrels or various other numbers of recoverable proved reserves, or there are suddenly vast new stores of "yet-to-be discovered" which will give us a few more years, I give the benefit of the doubt to the optimists and start with either of two quantities in the U.S.: 50 billion barrels or 100 billion barrels.

- **Tar sands oil**

  In Scenario 9, I start with 200 billion barrels to help answer the controversy about Canadian tar sands oil and the KXL pipeline. This proposed thirty-six inch pipe would facilitate delivery of some mix of Canadian tar-sand bitumen and highly-volatile Bakken "fracked" tight oil to the refineries in the gulf coast. **The KXL pipeline maximum daily-volume of 800,000 barrels would supply less than four-percent of all U.S. oil, or six-percent of the total of all liquid fuels used for transport.** At this rate, Figure 2, scenario 9 shows **an extension of the U.S. oil age to twenty-five years at the present rate of consumption with no other imports or exports and assuming the U.S. motorist can afford tar sands oil.** This minuscule band-aid for the U.S. oil addiction cannot possibly justify its incredible environmental impact and barely-positive EROEI (including massive amounts of diesel for giant trucks and excavators) of production. As this book is readied for printing in November 2015, TransCanada has placed on hold its application for the KXL pipeline. **This move effectively removes another 200 billion barrels of remaining world oil reserves thus bringing us beyond the peak of all oil, conventional and/or nonconventional.**

- **The importance of U.S. population change**

  For years I have been a strong advocate of population control as well as being a believer in peak oil. Part II Chapter 4 Figure 7 (World Oil Extraction and Population) shows the population growth and decline rates for various fertility rates of children per female (cpf) based on the older-demographic age-distribution profile of the U.S. **We now see exactly why a population momentum of 1 cpf (or even 0 cpf ... no more children) cannot possibly reduce population, in any closed-border (no immigration or emigration), society fast enough to keep pace with the declining Hubbert's curve for finite oil** ... even with the benefit of the doubt of 1.8 trillion barrels of recoverable oil remaining (1.6 trillion without Canadian tarsands). In Figure 2 only Scenarios 6 compared to 8; and 5 (the only scenario leading to a longer-than-one lifetime future) compared to 7, **extend the oil age long enough for the difference between one and two children per female to have an effect. So much for unpopular population talk.**

In round numbers, U.S. population is increasing about one percent per year. Four- million new births every year, plus one-million immigrants, minus two million deaths of senior Americans born when there were far fewer people starting families, adds up to a net increase of about three million Americans per year: **two hundred and fifty thousand per month, all expecting food, social welfare, a job, and a share of the American Dream. Within the time scales of seven to thirty years remaining in the U.S. gasoline age, the change in population, even at one child per female (1 cpf) will be far less a factor than physically running short of oil for all forms of transportation.** Closing our borders to any further immigration would still leave two million new Americans every year.

- **The effect of import or export of oil to the U.S.**

  This variable will do much to answer questions related to geopolitics, U.S. involvement in world crises, and U.S. fuel-export economics. Presently we import about half of our conventional crude oil (7½ Mb/d). We can do this only as long as residual (but waning) American wealth (as used for gasoline consumption) trumps the purchasing power of the rest of an energy constrained, food starved, and over-populated third world competing for what's left at less than 3 b/p/y per year. Should we export our crude oil or refined liquid products because a few advanced countries, or the wealthy minorities in third-world countries, can still outbid the poorer American consumer? How much longer will our global military presence (with its priority-consumption of 1 Mb/d of our waning oil endowment) be able to keep the world's remaining oil moving in our direction?

  Scenarios 2 and 4 are **in my opinion the most likely business-as-usual scenarios indicating twenty-eight and fourteen years left for our gasoline-intensive lifestyle**, including the mitigating effect of continuing to source one-half of our oil from foreign countries including Mexico and Canada. These would quantitatively be the same time-frames as if we suddenly reduced our per-capita consumption by one-half down to the Western European and Japanese level of 11 B/p/y, and continued to rely only on domestic oil with no import or export.

CONCLUSIONS

Clearly, there is no easy way forward

The curves and numbers in Figure 2 tell the story. The best we could possibly do, with heroic and highly-unlikely measures such as reducing per capita consumption from 22 b/p/y to 3 b/p/y, is shift the two-lifetime oil party towards the end of the eighty-year lifetime expected of a child born today. This is why I remain a strong, albeit unpopular, advocate for **not having children.** It's not that the extra consumption by increased population would make a significant difference in the short time frames of any of the various Scenarios in Figure 2 except for 6. Instead, it's the extreme challenges these children and their parents will face as the familiar American lifestyle crashes along with our complete dependence on gasoline, cars, travel, and all the supporting infrastructure (refer again to Appendix A).

The only possible window of opportunity for a U.S. long-term sustainable future is Scenario 5. This path combines the extraordinarily optimistic acquisition of 100 billion barrels of remaining domestic oil, a 3 b/p/y world average per capita consumption, a reduction of fertility to one child per female (1 cpf), and no immigration.

Scenario 7 shows the same goals, **but the oil-age ending in 90 years instead, because population at 2 cpf did not eventually decrease as with 1 cpf.** Population continues to increase and then stabilizes at "replacement" level as shown in Chapter 6. Scenarios 6 and 8 only give us one-half a lifetime left. There is not enough time for different fertility rates to make an impact.

Climate change and environmentalism divert attention

The American public is bombarded daily by **these perfectly legitimate concerns.** However, the time frame for serious impact is far longer than the eight to thirty year crises we face in the most likely Scenarios 1 through 4. **Also, it will be much easier (at least somewhat possible) to adapt to weather changes than to continue forward without the oil absolutely essential for every mode of transportation, our oil-intensive food system, access to the other two fossil fuels, all other energy sources including clean solar-electric, national security, and the thousands of other oil-based products which are critical to modern civilization.** I am a firm believer that mother nature will some day prevail after we run out of the earth's oil we used to devastate the ecosystem. Coal is a longer-term issue, but cannot be accessed or distributed without oil. And we've already extracted the easy, high-energy, cleaner grades.

Our rural farm in Maine was established long before the fossil-fueled industrial age and global warming. I can't imagine how our predecessors could have cleared the land, built stone walls, built our sturdy buildings, and had a long-enough growing season to grow 100-day beans, tomatoes, squash, and corn **if the weather was significantly colder then than it is now.** In my 40 years on this farm, I see little long-term evidence of global warming but, instead, colder winters because of arctic-melting-caused polar vortexes. It's not wise to talk to Mainers about global warming when we had our first frost in September 2014 and winter days below zero by December. December of 2015 and record breaking warmth have been exactly the opposite. But no one is complaining about lower heating bills.

Food will be the critical issue as we face collapse

The trade-off we imminently face is: is our continued addiction to gasoline more important than going hungry a few years hence? It is a well-established fact that we Americans are consuming about ten Kcalories (K=one-thousand calories) of **fossil-energy content** for every Kcalorie that we use for our personal food. Certainly, climate change exacerbates this predicament. But the fact remains that, thanks to fossil fuels (including natural gas-based nitrogen fertilizer), the current U.S. corn harvest is at record levels including supplying one-million barrels per day of ethanol and bio-diesel non-food yield with very poor EROEI. This synergism can not last without oil input as its backbone. (See Part III, Chapter 8 for more of my life-long farmer's perspective on this subject.)

A third of the world is already collapsing into food-crisis mode because regional populations over-shot the carrying capacity of their local support system. This is similar to nearly all previously-failed civilizations but, back then, on a regional basis. Now, this human predicament is playing out on a global scale, especially since declining national oil exports can no longer pacify ever-growing populations. This is a major context behind the riots and civil collapse in Egypt, Syria, Libya, and Nigeria. Even countries like Saudi Arabia and Iraq which still have surplus conventional oil have less and less to export each year after keeping their own growing populations fed and under control. Relentless depletion (over-pumping) of ground water adds to regional crises and refocuses the blame back to climate change and population.

Why does everything suddenly look so rosy in the U.S.?

Unemployment is falling. Home sales are improving. Car sales are up (to dealers and on $860 billion dollars credit). The stock market is skirting record highs. Gasoline prices are down. This short-term remission is a result of the resurgence of U.S. oil production from the 5 Mb/d level to above 8 Mb/d in the last five years. The remission is exacerbated by long-term "demand destruction" described earlier. Also, the dollar is stronger against the euro and yen because the other industrialized countries do not have a last gasp of their own oil endowment to revive their economies. Just because a terminal cancer patient sits up in bed or we have a warm stretch in October doesn't change the bigger picture. This grace period is the last chance to recognize and gain some control of our fate. It's too bad we didn't get serious in 1971 when U.S. oil peaked, or in 2005 when conventional oil extraction in the world leveled off. Instead, we continue to live for the day and totally ignore the future. This sounds like a "three-little-pig" story. Even squirrels and migrating birds know better.

How could Americans possibly get to 11 barrels per person per year, or better yet, 3 b/p/y like the rest of the world average? We will have no choice. It will happen whether we like it or not as we run short of oil. If we accept reality and start immediately we still could buy some time for a few more years and have some chance to mitigate our fate. I can only offer the following thoughts:

1. **Get involved! Help to exponentially network this story and book.** Distribute more copies. Reveal the American gasoline "elephant" on blog sites, letters to the editor, rallies, any thing you can think of. Your added energy to a movement is the only way to help. From the Chinese proverb: "To know and not act is to not know." The sub-heading in my fourth-edition manuscript was: "It's up to you." The source is not important. The unequivocal math is the message. I will make my book available at cost or free if necessary.
2. **Whenever a critical resource is in jeopardy, the only orderly recourse is equitable rationing.** Otherwise, infighting and wealth disparity (price rationing) leads to chaos. Increased fuel taxes, as in Europe, are regressive and the wealthy continue to consume while the poor walk. Everyone must equally share the necessary drastic reduction. There are myriad ways we could immediately reduce gasoline consumption and be better off as well **if only everyone else had to do the same. We all need to be forced out of our cars.** I would suggest that half of U.S. gasoline consumption is frivolous and not much more than an unnecessary joy ride, too fast, in a comfortable four-thousand pound chariot our ancestors could not have imagined. A controlled decrease in gasoline (and oil) consumption would divert billions of dollars elsewhere into the economy and revive the demand for other products which have been left at the curb. There are a million reasons why national, tradable fuel coupon (TFC) gas rationing can't be instituted and won't work. My only response is: what is the alternative?
3. An immediate fifty-percent reduction of our 9 Mb/d gasoline addiction would still leave U.S. oil consumption at 1.6 billion barrels per year just for gasoline. **This would slot the U.S. gasoline-consumption bloc still third in the world (tied with Japan's total oil consumption).** One would think that we Americans could make that "sacrifice."

4. Personal diesel use could be left un-rationed for a few more years. There is not time or wealth left for a mass exodus to more efficient small diesels as has been the norm in Europe because of much-higher fuel taxes. Hopefully the VW debacle will soon be resolved.

5. Extravagant use of commercial gasoline and diesel must also be curtailed. Eighteen-wheelers, school buses, UPS delivery of a mail-order economy, six-day a week mail delivery; all cannot continue. Possibly, with a national educational effort, each can be addressed and downsized in an orderly manner before our economy collapses into fuel-starved chaos.

6. The American love affair with gasoline-powered entertainment must soon end. Stock car racing, ATV's, RV camping, jumping into the car to drive 50 miles for a foot race or skiing are all totally unnecessary examples of our future being sacrificed for nothing but fun the today. Whoa! Now we're really getting serious. "Not my favorite pastimes"! The best I can answer is that with equitable rationing we could all make personal decisions to save, or barter open-marketable TFCs to continue some of our own life choices.

7. Legitimate commercial agricultural and critical public service needs would be exempt. Obviously, national defense needs of about 1 Mb/d today will take priority.

8. The administration of a TFC rationing system would be immense. I'm old enough to remember WWII gas rationing when my folks had to get by with three gallons per week. That was during a "National Emergency." Can anyone think of a greater emergency than we face now with seven to thirty years left (Scenarios 1 through 4 in Figure 2) in the U.S. oil age? Remember, our only last chance for mitigation is to start now.

9. National gas rationing would give immediate impetus to myriad transportation alternatives. Electric cars, bicycles, public transportation, electric tractors for small farms, and healthy walking would all become popular **if there is still oil left for the absolutely-necessary peripheral support of all non-oil transport.** Energy-intensive sprawl would be discounted in value. Localized food production and non-travel entertainment and healthier lifestles would be enhanced.

10. Air travel would remain the domain of the diminishing percentage of Americans who can still afford to fly. By drastically reducing profligate gasoline consumption the other means of public transportation and the traditional American lifestyle could continue a few more years ... possibly into a far-distant sustainable future if Scenario number 5 finally combines

dwindling oil with a significant reduction of population. Our children could at least have a taste of the fading lifestyle we took for granted while burning away their energy inheritance.

11. Regional home heating with oil can continue only a few more years as the attention focused on finite oil would encourage the transition to better efficiency, bio fuels, and solar. Grid electricity from coal, nuclear and hydro. will last longer than oil, but all require oil to function, another reason to keep the oil-age going albeit at a much lower per capita rate.
12. See Part III Chapter 7 for further discussion and quantitative details regarding gasoline rationing.

WHO AM I AND WHY DO I CONTINUE THIS UNPLEASANT WORK?

No one else seems to be telling the whole story especially with the daily bombardment of hopefully-self-fulfilling prophetic, feel-good news, cherry-picked partial solutions, and copious misinformation. **The quantitative interaction between U.S. gasoline consumption and the world oil market is unreported anywhere else.** The "experts," from our politicians, to economists, to the investors, to most everyone associated with energy and oil; have a nearly unanimous intent to perpetuate our high-energy lifestyle and cannot comprehend the pace that we're running out of gas. Even the population experts and ngo's infer that two children per female (cpf) is an acceptable goal; if and when reached, "problem solved."

As a retired R&D engineer, I come from a business career and mind-set dedicated to pondering the future and patiently directing the back-room development of the next-generation product line. This, while the company is at the top of its game. My goal was to be ready with the answers **before the company president calls a crisis-mode meeting.** I'm also a "numbers-guy" attempting to find the facts and quantitatively see through and around much of the subjective obfuscation we hear every day. Finally, as a life-long farmer, my ancestors were ten generations of "hard-rock" New England farmers. Where and how does food-energy come from? I vehemently support the re-localization, solar-electric (with respect to battery limitations), permaculture, resilience, and healthy food movements. That's a primary reason we moved to Maine 40 years ago. We grow and store a large share of our vegetarian diet. Unfortunately, these efforts on personal and local levels do nothing to address the imminent, easy-energy, gasoline-fueled crisis. The technological "magic" of solar cells and modern electronics is well established, but their future potential can't possibly work without continued minimal support from remaining liquid fossil fuels of at least 3 b/p/y.

PART II

SUPPORTING INFORMATION

Moving beyond a basic understanding of U.S. dominance in world oil markets, we we must accurately identify our final options on a macro basis as well as personal preperation for a vastly different future.

The post-oil age looms large and complex. There is much obfuscation about the potential and limitations of finite fuels of all types. A transition to sustainability and/or renewable forms of energy is our only choice. There is no chance of seamlessly substituting dilute annual sun-light energy for millions of years of conveniently stored fossil fuels.

CHAPTER 3

Education and Personal Action

I have at least 300 books in my library that relate to, or directly address the end of fossil energy. Most are listed in the bibliography at the end of this book and several of the most pertinent are cited throughout the following chapters. **Only a few broadly connect the peak-energy-food-economic subjects with the steady growth of population and demand.** The classic references on the role of population in resource overshoot and societal collapse: Meadows, Bligh, Bartlett, Carr-Saunders, Diamond, Erlich, Grant, Hardin, Ruppert, Tainter, and more; have not made a meaningful impact on the main stream public. **The messages are too dire and there are many, many other voices offering solutions that are much more palatable.** Especially disconcerting is the conventional wisdom that population must continue to grow to provide economic growth and social security support for the elderly. For instance, in "USA Today"(2/21/2013) is an article with the heading: *Lower U.S. Birthrates a threat to our future, Lawmakers need to promote families.*

In addition, nearly all demographers perpetuate the conventional wisdom that a fertility rate of two children per female (2 cpf) is a"replacement rate" and a satisfactory goal for long-term survival. The methodology introduced in Chapter 6 of this book mathematically shows the fallacy of this reasoning when modern health care and old age are included.

A third group of vociferous activists (and publications) concerned with environmental and climate concerns rarely discuss peak oil and population. The best studies that combine all three subjects: energy, population, and environmental degradation; and their role in previous failures of civilizations (Tainter, Diamond, etc.), are relegated to the back lower shelves of book stores, out of mainstream view, conversation, and interest. Finally, an understanding of this "triple crisis" is completely missed and/or debunked by the classical views of economists.

Now, time is running out and we are faced with at least three major obstacles to any hope of perpetuating a vestige of our modern lifestyle:

1. The mainstream public is not hearing the complete and accurate story. A threatening hurricane, tsunami, or pandemic would be all over the news. **But the greatest challenge ever, the nexus between growing population and declining energy, both related to climate change, goes barely noticed.** The environmental and climate-change advocates seem to be losing ground with the American public, especially when economic growth and job creation are argued in opposition. **The energy-industry barrage of TV ads to convince the public that "there's plenty left" far out weighs the Cassandras who are more concerned about the future of modern civilization.**

2. **Many of the tiny (5 to 10%?) segment of the U.S. and world public who do understand and believe in the imminent decline of energy have chosen to retreat to the sidelines.** This may be because of hopelessness or just human nature to avoid the emotional, physical, and financial investments required to attempt to make a difference. **Inaction by those (you?) who understand the energy vs. population predicament will only facilitate the demise of modern civilization.**

   For the last ten years I have been writing and speaking to generate public concern about the imminent decline of fossil energy. In that time, the combination of population growth, environmental devastation, food shortfalls, fuel cost gyrations, and periodic economic recessions have exacerbated the human predicament throughout the world. The internet is rife with every aspect of these ominous rumblings, but civilization just keeps groping forward with no direction or concern for the future.

   As improved extraction technology and more non-conventional liquid fuels were made possible by higher prices, we in the U.S. enjoyed ten more years (days of grace) of our high-energy lifestyle. However, in this same time-span, dwindling residual wealth and growing debt accelerated the disparity between the remaining few who can outbid the growing poor. This last remission from our terminal illness will be totally wasted as we doggedly attempt to continue business as usual. It is probably only a dream that the concepts proposed in my humble effort, plus similarly alarmed voices, will make enough impact to give ourselves and our descendants hope for a future with the life we've taken for granted. At least, suggestions offered for personal preparations may help a few readers.

3. The escalating tension (widening gap) between declining energy and expanding population may be so far established that no long-term solution is possible. **The best we could possibly hope for is to drastically reduce per capita consumption in the U.S. to, at most, 3 barrels per person per year (b/p/y) and simultaneously reverse population growth as quantified in Chapter 2, Figure 2, Scenario 6. These drastic moves would significantly extend the fossil-fuel era for a few more generations and support a nascent, much lower-energy future.** This would be in sharp contrast to the impending shock of overshoot, collapse, and mass chaos. The children born today and for the next few years could still have a small chance for the remnants of the modern high-energy lifestyle we took for granted.

Imagine the doctor coming from the hospital laboratory to the anxious patient in the waiting room: **"The bad news is you have a terminal illness and have only six months to live. The good news is that if you do exactly as I say, and drastically change your diet, lifestyle, and habits; you might, instead, have six years and possibly many more."** Most of the public would rather look for a comforting second opinion and continue without change. **The difference between this personal scenario and the imminent, world, energy-population macro-crisis is that typical individual response and inaction will take every one down together.**

Also, our problems are not limited to energy, population, and the environment. Continuing this logic path, I will quote from page 83 of my 2006 book, "The End of Fossil Energy":

> Eventually, even the ore for each material will become so depleted that all subsequent material requirements will have to come from recycling. By that time we will experience population reduction and drift slowly backwards from any semblance of an industrialized modern civilization. It may be that the 200 year fossil-fuel consumption curve is just a small blip in a much larger ten to twenty-thousand year finite-metal- resource epoch that spans the copper, aluminum, and iron ages of civilization. It could be argued that it is impossible for an advanced civilization to exist anywhere in the universe for longer than a few thousand years. The very unique circumstances and time required to concentrate metals and energy on a singular planet cannot provide enough resources to last longer than a few years in the cosmological time frame.
>
> The time frame for space-travel may, in reality, be impossible. This is not a unique concept and is discussed further in Chapter 4. (Also see, *The Olduvai Theory,* greatchange.org)

To summarize again, there is absolutely no chance for a repeat of the fossil-fuel energy age in less than the next few hundred-million years, the time to store new concentrations of photosynthesized sunlight. If even that were to happen, our future descendants would still have to find depleted ores or use scrap remains of structural metals. By then, the earth's climate may be too altered for human life as we knew it.

As I wander among the stonewalls in rural Maine, I like to think that, about 200 years ago, the first settlers had enough food and surplus energy to clear the forests of wood and the fields of prodigious numbers of rocks and boulders. At least, back then, there was plentiful iron ore to make guns, axes, saws and stone boats for the oxen. Several centuries into the future, will our surviving descendants again come to equilibrium with the steady-state carrying capacity of the immediate surroundings, with or without the ores for structural metals and enough energy for smelting? **They, also might ponder the silent walls of stone intermixed with the remains of huge buildings and high-speed overpasses, and wonder what our life was like in the brief fossil-energy epoch**.

I also have a dream, that if we were smart enough to do everything just right, and planned enough time to prepare; we could, on a limited scale, and a regional to national level, **transition to a limited, solar-electric powered, high-tech society**. With minimal remaining oil and far fewer people constrained by low-energy travel to a smaller locality, could we carefully nurture a semi-modern sustainable lifestyle for generations to come? **This could happen with a mix of PV, wind, solar thermal, and hydro-power electricity.** We might even grow limited biofuels and lubricants in lieu of critical food to keep our machines rotating. For this to be remotely possible and still have concentrated energy necessary for personal transportation and agriculture, we would still have to resolve the battery-storage-recycling problem discussed in Chapter 5. And, of course, we'll have to muster the energy and manpower necessary to defend the borders of this unique surviving utopia. But of course, none of this will be possible unless we first significantly reduce per capita energy and population.

From page 129 of my same book, referenced on the previous page:

> In conclusion, I offer a quotation from a lecture series titled, "Of Men and Galaxies" given (then) 39 years ago in 1964 at the University of Washington by cosmologist Sir Fred Hoyle:
>
> *It is often said that, if the human species fails to make a go of it here on earth, some other species will take over the running.* ***In the sense of developing high intelligence this is not correct.*** *We have, or soon will*

> *have exhausted the necessary physical prerequisites so far as this planet is concerned. With coal gone, oil gone, high-grade metallic ores gone, no species however competent can make the long climb from primitive conditions to high-level technology. This is a one shot affair. If we fail, this planetary system fails as far as intelligence is concerned. The same is true of other planetary systems. On each of them there will be one chance, and one chance only.*

Assuming you've read this far, by now you have begun to grasp the enormity of the interrelated synergism of finite energy (especially peak oil), food for a growing population, environmental degradation, egregious American liquid fuel consumption, and economic collapse. You will probably also need to work-through the expected stages of reaction to serious news: denial, depression, acceptance and now the best response, becoming proactive. **How can you, as a concerned individual, make a difference and improve our chances?** If nothing else, becoming involved will provide a good feeling of action, and meeting others with the same mission. Having thought about this question for many hours and years, I will offer **my best suggestions for immediate personal involvement;** in some order of importance, although all are absolutely necessary.

PERSONAL ACTIONS

If you have not already done so, immediately begin to plan serious food and energy independence for you and your immediate loved ones. A crack in the fragile world house-of-cards can occur at any time, maybe in 2016 as an Iranian confrontation, growing terrorism, another Arabian Spring spreading to Saudi Arabia, or the collapse of the European Union. **The music could stop at any minute.** Watch the news and the price of oil, up or down. In any event, **don't be too far from your personal chair.** Don't plan on running to the bank or calling your broker. Rest assured, no one will come and rescue you when everyone is diving for cover. You will be on your own. The electrical grid, a natural gas pipeline, food in the grocery store, money in the bank, and/or fuel at the gas station may suddenly become unavailable. **The following suggestions are offered to help you prepare for the coming tsunami:**

Start your own garden

Read Chapter 8. Find at least 200 square feet of soil and sunlight. Get down and dirty. There are thousands of grow-your-own books, websites, and magazines. Start with the permaculture movement. Join similar localization activity. Phase out the lawn and even the flowers. Grow things you can eat. Learn about diet, health, storage, pests, open-pollinated seeds, calories, beans, and other plant proteins. The next step might be to get a few chickens for eggs. This will be fun and exciting. You will be joining millions who have seriously started this pro-active path and are already healthier for it. The "Victory Garden" movement in WW II is a precursor.

Build your own stand-alone PV solar survival system

No grid tie or generator is required. Minimal requirements are one or two 100 watt PV panels, charge controller, a 40 amp surge protector (fuse), and at least (4) deep-cycle 12v storage batteries connected in parallel with about 100 ampere hours of storage in each. This will provide about two kilowatt hours of storage at 50% depth of discharge to preserve battery life, and surge capacity for lights (LEDs!), refrigeration, entertainment, and communication, **but no heating.** Appropriate meters for incoming and outgoing current (amperes), and system voltage will show what's happening. Our ancestors would have been amazed. A 1000 watt (1500 watt surge) 12 volt DC to 120 volt AC inverter will provide minimal power for 120 volt requirements. Total retail cost: panels, $300.00, batteries $400.00, inverter and charge controller, $250.00, everything else, $250.00. Altogether, not much more than $1200.00. Much more on this subject, including battery limitations, are covered in detail in Chapter 5, "A Solar-electric Future."

Be self-sufficient

It's always common sense to have a year's supply of food, especially non-perishable bottled, canned, and dried; to supplement the produce from your new garden. Don't forget seasonings, honey, tea, and coffee. With a solar-powered freezer, the surplus from your garden can be stored for years. Accumulate as much as possible of dried goods, paper and cleaning products, matches, candles, propane bottles, bullets, and (lots of) batteries of all sizes. Necessities may become more valuable than worthless paper money. Don't forget books, games, sports, arts, crafts, and music.

Prepare for domestic heat and hot water

Have at least several weeks of potable water available. Install a domestic water system that will collect rain water or use your survival electrical system for pumping ground water. Keeping warm will be a huge challenge. The brief resurgence of "fracked" natural gas will suffice only for electric utilities and buildings fortunate enough to be on a functioning pipeline. Our sudden "one-hundred year supply" of "fracked" natural gas is totally impossible. You should prepare for a low-energy future by significantly downsizing your heated area and all water systems to avoid freezing. Heavily insulate a much smaller living area, not the whole house. Install solar hot water and solar-thermal heating to capture and store as much energy as possible. Accumulate plenty of winter clothes and blankets. Plan on a primary or back-up heating system that is independent from fossil fuels or an electrical grid. Firewood is still the best defense and can be stored for years. You can't make your own pellets, and it would require too much precious PV electricity to run a pellet stove. For a cost comparison; at $3.50/gallon, fuel oil provides 35,000 btu per dollar. Dry firewood for $200.00/cord gives twice as much value, or 70,000 btu/dollar. Electric heat is 100% efficient, but at $0.15/kwh will yield only 23,000 btu/dollar. Natural gas at $10.00 per 1000 cubic feet gives about 80,000 btu per dollar. Minimal air conditioning is more feasible in the summer when long days can provide the necessary PV power.

Most important of all, get involved and forward this story

Personally project, in every way possible, the facts and urgency of the developing population/energy crisis. This theme is repeated throughout this book and is a most serious call to action. **If we begin immediately, and do not break the communication chain, we can use exponential networking to reach vast numbers in a very short time. The versatility and ease of electronic communication might help but only if we use it**. My library on energy, population, and environment is huge but gathers dust. These are not new subjects. Now, however, the convergence of science, numbers, and time is knocking at the door. **We can give in to fast-unfolding history, or not give up and still try to make a difference. Why not? We have nothing to lose but our, and our kid's future.**

THIS BOOK, THE SEQUEL TO TWO MANUSCRIPTS

I, a humble author, retiree, and activist, am only a messenger with a disturbing crystal ball and comfortable familiarity with numbers, farming, and energy. As stated before, my own calculations and my library of books are the foundation for my call to action. There are many titles (and web sites) but **only the choir is reading the score.** The traditional channels of publishers and retail stores, or Amazon and the web are not making an impact. E-books are only ordered by those already involved. The internet is hopelessly overwhelmed with every blog-theme and counter-theme imaginable and we're buried in printouts.

My 2014 plan was to mail an 8" by 10" spiral-bound "4th Edition" manuscript, in hard form, directly to recipients as long as I could afford the time and money.

No publishers or distributors were needed. **I attempted to condense everything related to the energy-population-environmental triple crisis into one self-published document.** It was divided into a preface and nine loose chapters which may have been redundant; but each, or the entire manuscript could be reproduced locally. This Chapter 3 was included to explain the proposed ground rules necessary to reach a vast number as quickly as possible. We must break out of traditional channels of communication.

The power of the exponent

To reverse the direction of our energy-intensive dead-end trip will, in itself, require a huge amount of human energy; the sum of a little bit from each of many. **If this scheme has any chance of expansion, you the reader, must make it happen and keep it moving.**

In the 4th edition in a manuscript form, I proposed the following:

1. First, the recipient could add his/her own input, comments, name, and new contact information.
2. Single chapters, or the whole spiral-bound manuscript could be locally reproduced. **Chapter 3 should be included in entirety to explain the basic forwarding scheme and the need for exponential projection.** Each two-page sheet, black and white, costs about ten cents at a local print shop. The entire one-hundred pages (fifty sheets) plus spiral binding and a glossy cover should cost not more than ten dollars. The media -mail postage including packaging cost is another three dollars.
3. To start, I printed 350 copies and mailed them directly to my own list.
4. **If you, or any one else who wants to become involved in this last-ditch effort, copies of this final 5th edition book can be ordered from Amazon for $15. Or, I will send a copy directly to you as long as I can keep up. Just send your contact information, including address, to the email on the cover letter.** I will acknowledge your request, preferably by your e-mail address and take care of your order ASAP. I don't plan to make a cent but can only do so much on a retiree's budget and time. I will find help as needed as this effort gets moving.

5. Obviously, if there is any chance for the sheer numbers of this movement to get rolling (and I have to believe it can, it's our only hope), new reproduction and mailings must be done downstream. This is a grass roots movement that must be self-perpetuating.

6. **If each recipient were to forward just ten copies of the manuscript or book within a month, and only five of the ten recipients continued the chain on each following month, on the twelfth month 244,140,625 copies would be mailed. The total sent out in one year would be over three-hundred million! Get involved. Join the movement. Please, don't let down everyone who does make the effort. Your descendants will thank you.**

CHAPTER 4

In-depth Review: Finite Fuels, Renewables, all Resources

The intent of this chapter is to give a quick overview and update of all energy sources going into 2015. It is difficult to avoid becoming mired in the details of each of these subjects and risk losing sight of the overall basic conflict between the limits of finite fossil fuels (especially oil) and steadily increasing population. This is the foundation for the ultimate crisis that confronts all of civilization. To add to the confusion and complexity about energy there are many differing references, units of measurement, grades, and use-efficiencies for all forms of energy. To keep things as simple as possible; wherever possible, all units will be converted to energy-equivalent barrels of oil.

WORLD ENERGY SUMMARY 2014

Table 4 summarizes the contribution from all primary energy sources. These numbers fluctuate from year to year and there may be minor disagreements, but these will not change the conclusions. See also Figures 3 and 4 for graphics. **Note: nuclear, hydro-electric, solar, and wind are presented at three times their electrical energy equivalents in order to compare directly with equivalent fossil fuels burned at an average of 33% efficiency.** Biofuels are also burned at 33% efficiency. Solar and wind contribute less than one-half percent of total energy but at much closer (comparatively) to 100% efficiency. In other words, it requires only one-third as much electrical output at nearly 100% efficiency to offset fossil fuel combustion. These are examples of the confusion that creeps into any comparison between different forms of energy.

TABLE 4 Comparison of Energy Forms

Energy source	Equivalent billion barrels of oil per year	Percentage
Oil and all liquids	32 *(see also Chapter 1)*	36
Coal	24.5	27
Natural gas (w/o NGL's)	19.5	22
Total fossil fuels	76	84
Nuclear	5	5.6
Total finite (non-renewable)	81	90
Renewables		
Hydro-electric	6	6.6
Solar and wind	0.6	0.7
Biofuels (and everything else)	2.4	2.7
TOTAL	90	100

ALL LIQUID FOSSIL FUELS

Figure 3 shows that world extraction of **conventional** oil continued into 2015 along an "undulating" plateau of about 75 million barrels per day (27 billion barrels per year). Beginning in about 2000, the addition of **non-conventional** oil including tar sands, deep-water, polar, natural gas liquids, condensates, and "tight-oil" made possible by horizontal hydraulic fracturing, increased world output of "all liquids" by another 16 million barrels per day.

These recent additions, especially biofuels, have much poorer energy (and economic) returns on investment EROEI); some as low as 1:1 instead of up to 100:1 with original conventional oil. To include them together is like adding pears to apples, but they are shown this way in Figures 3 and 4. "All liquids" will help extend the oil age a few more years, but totally confuse the public by suggesting unlimited potential for extending industrialized civilization beyond the two-lifetime, oil-age epoch we are now half-way through.

Oil, 2014 update, history, and forecasts

It is nearly impossible to precisely quantify and compare all world energy statistics because of poor data and starting assumptions. There are too many different country, political, and economic viewpoints involved to avoid bias. For instance, Figure 5 shows a more recent International Energy Agency's (IEA) *World Energy Outlook 2012* forecast. This report says, "The world could produce **increasing** amounts of

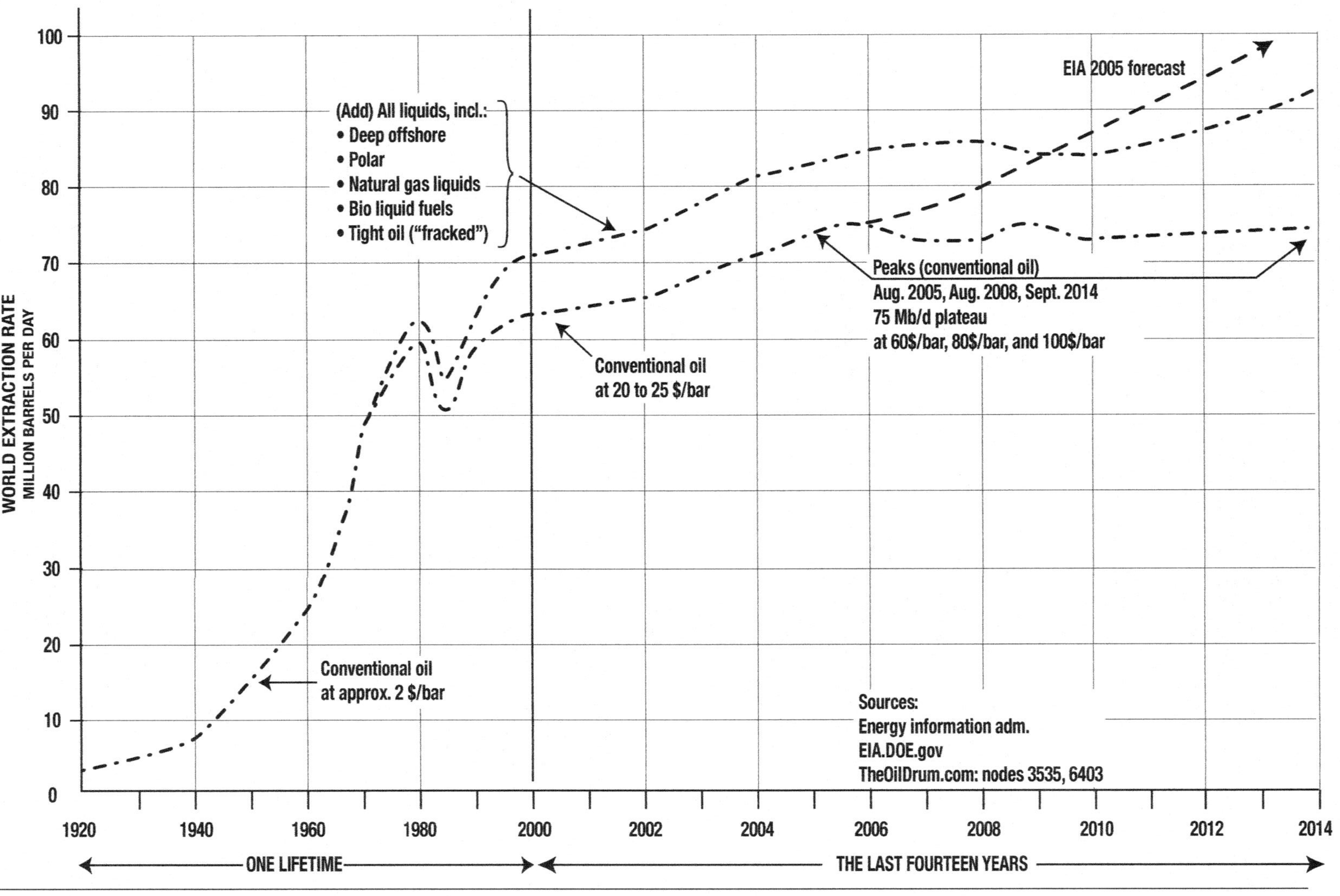

FIGURE 3 Peak oil update (January 2015)

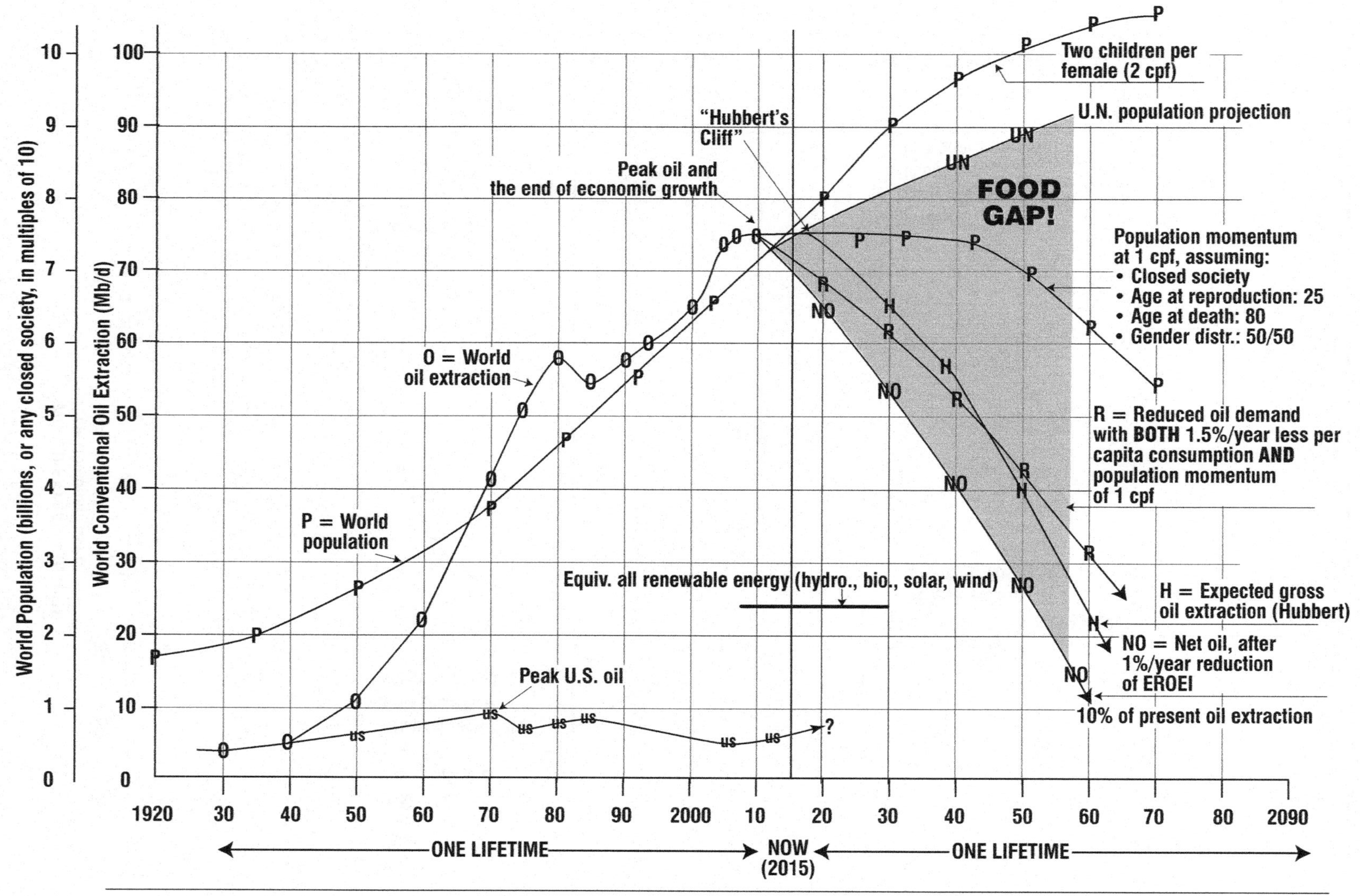

FIGURE 4 World Oil and Population in a Two Lifetime Span

oil through 2035 and meet the world's growing demand for energy as oil." Where does the IEA optimism come from? What happens after 2035?

The top-most curve in Figure 5 (AL) shows the IEA projection for all liquids, including non-conventional like tar sands, polar, deep off-shore oil, natural gas liquids, and condensates to remain essentially level at 33 billion barrels per year (90 mb/d) for the next 20 years. The lower level curve (O) at about 30 billion barrels per year (81 Mb/d) is for conventional oil plus natural gas liquids. **These IEA predictions do not include "yet-to-be-found" liquid hydrocarbons and therefore are flat. The U.S. Energy Information Administration (EIA) projections shown in Figure 6 show a steady increase for all three fossil fuels until 2035.** These EIA forecasts include "yet-to-be-found oil, and apparently refuse to acknowledge the steady decline (beyond peak) of output for more from half of the world's producers.

The two agencies seem to be similar, or are trying to out do each other in confusion and optimism. None of their rosy predictions seem concerned about the concurrent, steady growth of population which, by 2035, will reach about 8.5 billion at the present fertility rate of almost two children per female (2 cpf). (See Figures 4 and 7). **Neither of these agencies offers a clue about the future after 2035** for a continually-expanding population which has become precariously dependent on a fossil-fuel-based food system, and a world all-liquids energy supply of one billion barrels every eleven days. **Sooner or later energy, even in the most-optimistic crystal ball, will no longer be adequate and civilization will have lost another critical twenty years that might have been the last opportunity to stem the tide. By then, a child born today will be an adult.**

Also shown in Figure 5 are, in my opinion, the most realistic and optimistic projections for conventional oil (H = Hubbert's Curve), and a second curve (H+500) showing the effect of an additional 500 billion barrels. This scenario would increase the world total, original, conventional oil endowment to 2.8 trillion barrels, considerably more than the original Hubbert's and Association for Peak Oil (ASPO) predictions of 2.0 trillion barrels. In my methodology, I exclude "all liquids" because it seems incorrect to add more recent types of poor EROEI fuels to the baseline of conventional oil.

"Saudi America"

Also shown at the very bottom of Figure 5 (U.S.) is the glowing IEA 2012 forecast for a boom in U.S. oil. By 2020 and led by "tight oil fracking," U.S. extraction is projected to once-again reach its 1970 Hubbert-predicted peak of 9mb/d This renaissance began in 2010 as an increase of 0.22 billion barrels per year (600,000

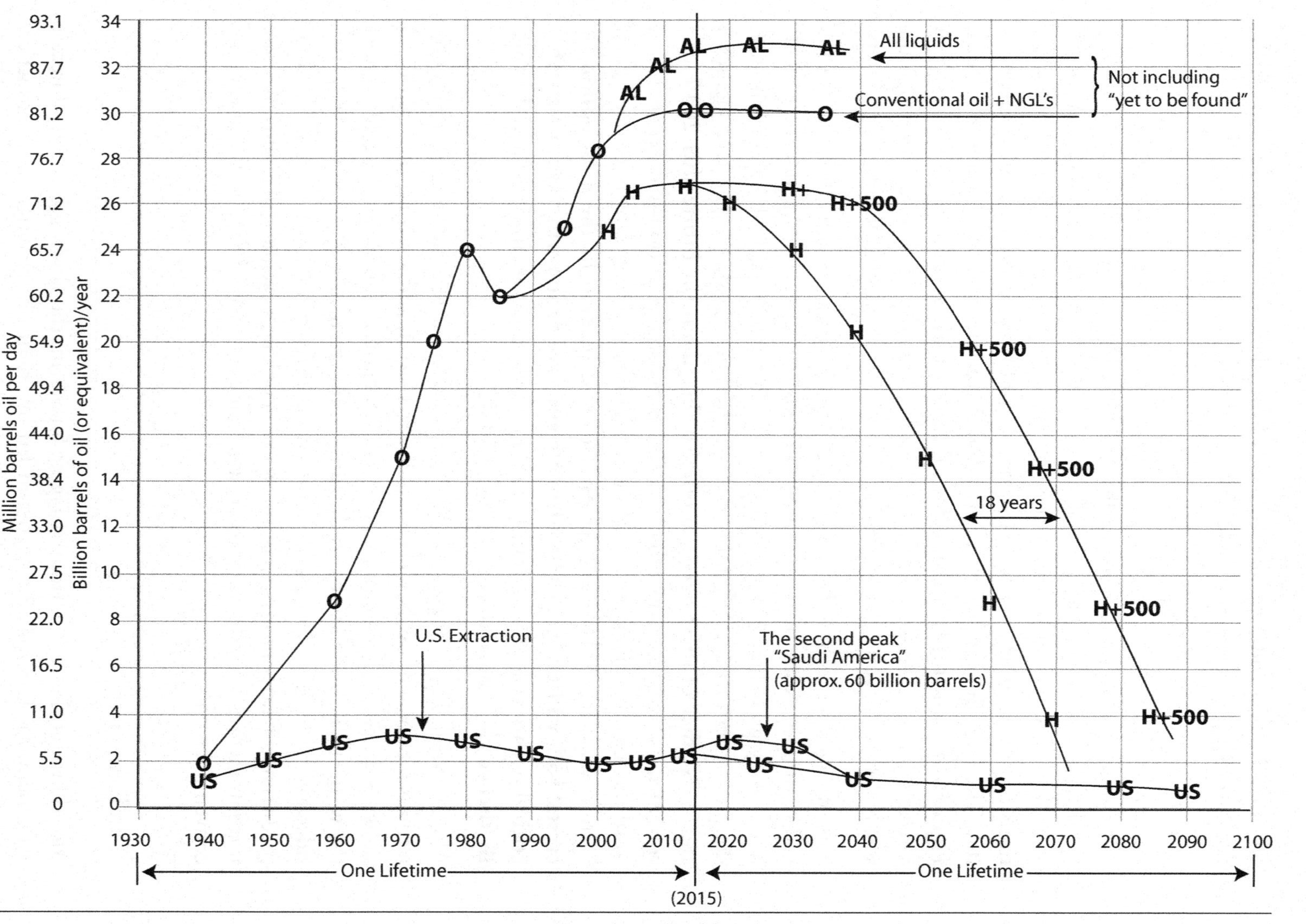

FIGURE 5 2014 Update, World and U.S. Oil

barrels/day). It is enticing to simply extrapolate this curve upwards another 1.5 billion barrels per year to a second, "twin peak" of U.S. extraction at 3.5 billion barrels per year, exactly what Americans are using each year now, just for gasoline. If this did come to pass, it would add an additional 25 billion barrels to the 75 billion barrels, the most we could possibly have left in the next thirty years. **The total of 100 billion barrels of remaining U.S. oil is the same as in the most optimistic scenarios in Chapter 2.** This euphoria assumes a massive amount of new infrastructure, new jobs, and predictions of a revitalized U.S. economy.

The additional oil, shown in Figure 5 as a "second peak," would add just over four years at our present, U.S. annual consumption rate of 7 billion barrels per year. This new extra oil would still have to accommodate continued population growth, even at one child per female (see Figure 7). **This surge of American oil, if it does come to pass, would all be over in less than half the lifetime of a child born today leaving twenty-five billion barrels less for their future.** How can anyone deny or argue this simple logic?

Coincidentally, by 2035, I will personally have reached one-hundred years old so probably won't be here to see the outcome. The oil age will have seriously contracted by then. In my opinion, we will know the outcome far sooner, in the next several years, because the predictions for world and U.S. oil extraction are far too optimistic in light of steadily decreasing EROEI and global net export math. (GNE, see below).

Other forecasts

In this amazing age of information, it is possible to go on-line and "google" any subject. "Peak oil" will turn up hundreds of references and forecasts, most within the ranges shown in Figure 5. Typical of most economic pundits, Larry Kudlow ("The Kudlow Report") assures us that "we are becoming energy independent." To this euphoria, I would add: this includes coal and natural gas for a few more years, or as long as we can extract our kid's future energy as fast as we can to maximize immediate profits and continue business as usual. Some analysts are even more optimistic. For happier reading try: *The new Era of Oil Renaissance* (commodities-now.com/13445) or, *The New American Oil Boom* (secureenergy.org).

The summary curves shown in Figure 6 are from an excellent, comprehensive, peer-reviewed paper by G. Maggio and G. Cacciola titled: *When Will Oil, Natural Gas, and Coal Peak?* It was published in the International journal, *Fuel*, and can be found on www.sciencedirect or wikipedia. **This paper projects oil extraction to decline fifty-percent, from a 2015 peak of 30 (82 Mb/d), to15 billion barrels per year (41 Mb/d) by 2050.** This about the same decline as the Hubbert prediction shown

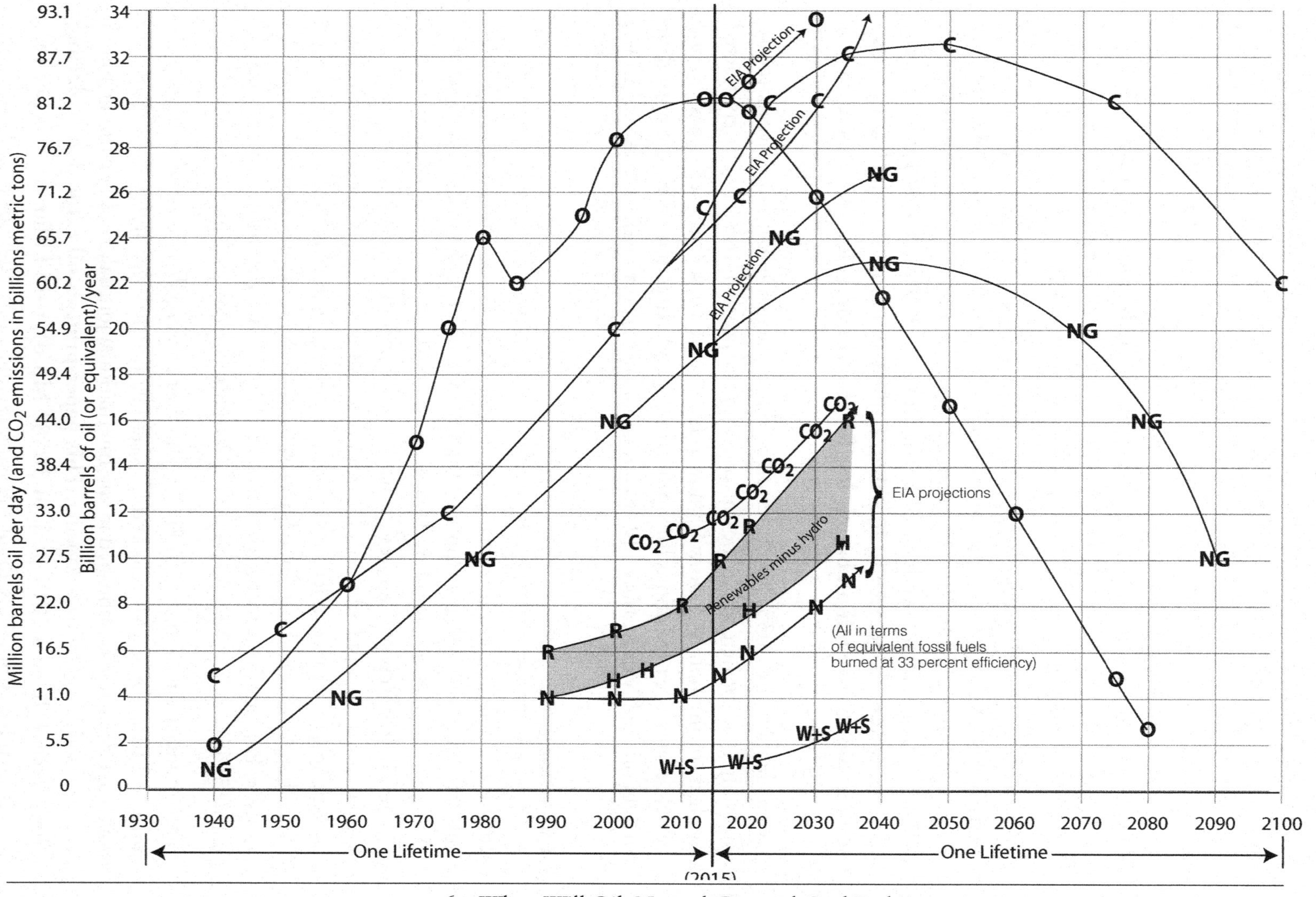

FIGURE 6 When Will Oil, Natural Gas, and Coal Peak?*

* per G. Maggio analysis and EIA projections *(source: World Energy Consumption Outlook)*

by the "H's" in Figures 5 and 7, but with the addition of NGL's. *The Maggio* paper also shows peaking and decline curves for various "ultimate" endowments of all fossil fuels. These are the mid-range reference cases for each fossil fuel.

The initial world-endowment for oil shown in Figure 6 is 2600 billion barrels, just slightly less than my most-optimistic scenario of 2800 billion barrels shown in Figure 5. **The ominous, predicted declines in oil, no matter how optimistic, are blithely ignored as we roll merrily along, pedal to the metal, with steadily increasing numbers of consumers, all looking forward to economic growth, travel, jobs, fuel, and food.** We are comforted by the steady upward trends shown in Figure 6, from the EIA.DOE World Energy Outlook. The optimists are quick to include "yet -to-be-discovered" to make their case.

Global Net Exports (GNE)

We can not leave an update of the oil age without including another very serious, overlooked issue that is not considered in "plateaued" world extraction rates. **There is steadily less net global oil left available for export** from traditional oil-rich countries: like Saudi Arabia, Indonesia, UK, and Venezuela, **after they first satisfy growing internal demand.** As autonomous nations struggle to supply their increasing national oil demands; the overall, world oil export numbers steadily contract, especially so when considered on a per capita basis. The poorly understood phenomenon of GNE underlies and explains much of the political tension and economic distress in the world. For instance, Indonesia was one of the original OPEC countries (and supplied much of Japan's oil in WWII), yet ceased to be an exporter by 2003. The UK pumped out so much of its North Sea endowment during the heady Thatcher era that, after peaking in 1999, it ceased to be an exporter by 2005.

Another excellent example of terminated net exports is Egypt, which recently ceased being an oil-rich exporter in 2010. Two simple exponential curves, one of increasing population, the other of declining oil extraction, quickly intersect. **The results are crowds of unhappy young protesters for whom the oil party just ended. This was the context for the "Arab Spring" and continues to this day throughout much of the world. Geopolitics is nearly always related to oil. Any form of government will be unpopular if it can no longer pacify a growing population with declining energy income.** This is a classic example of "more forks and less pie." Neither a Democracy nor Autocracy can change the numbers and inevitable outcome. Some countries, like the UK, still rich in residual wealth or other resources can weather this growing storm for a while longer, but eventually all must face the decline of oil and an unhappy populace.

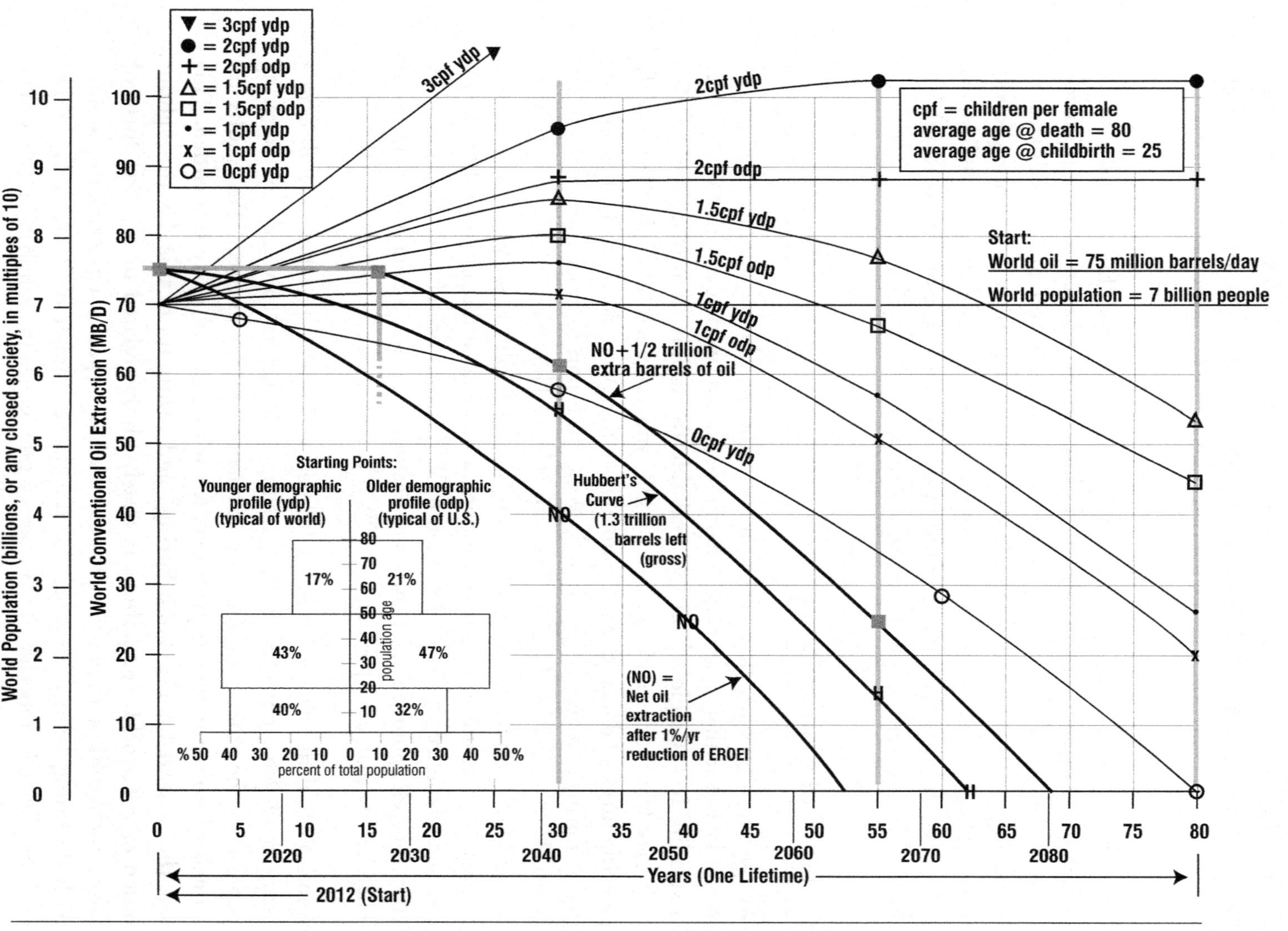

FIGURE 7 World Extraction and Population (beginning in 2012)

A complete explanation of the phenomenon of GNE has been developed by Jeff Brown of ASPO-USA(.org). Using the same methodology that predicted the end of oil exportation for Indonesia, the UK, and Egypt, Jeff projects that "Saudi Arabia will cease to be a net exporter in 2031" (ASPO-USA correspondence 4/11/2012). Coincidentally, this is about the same end of time frame as the IEA and EIA projections to 2035 as shown in Figures 5 and 6. There are no more optimistic projections of an oil age beyond then, one generation from now.

The U.S. has long since ceased being a net exporter of oil. The projected second domestic "oil boom" from an extraction rate of 5, back to 9 Mb/d, even if comes to pass for a few years, will not reach our present (with no concern for increasing population) consumption rate of nineteen million barrels per day.

To summarize, it is inevitable that the ratio of diminishing oil availability for export, to that left to satisfy the increasing demands of importing nations, will rapidly decline as more exporting countries divert their remaining endowment to keep their own people happy. Yet, there are still a few politicians advocating that we export some of our "Saudi America" oil for short term profit, even as we continue to import the difference between our 19 Mb/d consumption and 10 Mb/d extraction.

NATURAL GAS

Natural gas **can never** replace oil as the fundamental finite fuel for modern civilization. Despite rosy statements by "experts" like T. Boone Pickens, **natural gas is not a large-scale transportation fuel.** Consider the basic physics: All forms of motive power (except for dilute, sporadic, wind or direct on-board solar) **require that the energy source (fuel) be stored, and carried along for the ride, until the next refueling opportunity**. Weight, energy-density, and volume of the fuel, or any other on-board energy storage system (like a battery, bale of hay, or back pack of food) are critical. The movement of anything of significant mass, especially a truck or an airplane, requires considerable energy to do the work (force times distance). This is why liquid fossil fuels are used for 90% of motive power and coal is a poor second choice (see Chapter 7, "Gas Rationing" for more discussion). Compressed natural gas (CNG) is minimally used around the world for short-distance cars and trucks; **but this is only possible where the supply and infrastructure are readily available for handling a highly flammable fuel at 3000 psi.** CNG should not be confused with heavier, natural gas liquids (NGL), primarily propane, at 300 psi. Propane is heavier with three carbon atoms, instead of one, per molecule like CNG, and is a most desirable by-product of the current natural gas boom. The recent surge in NGL's is a primary reason for the temporary remission of the oil-age, but at best, propane can add up to seven million barrels per day of additional liquid fuel (about nine percent of world conventional oil extraction). A propane-based transportation system would be highly dangerous and require entirely new infrastructure and vehicles. Airplanes and 18-wheelers would not fit into this paradigm.

Oil, with more than six carbon atoms per molecule, is heavier. It has an energy density at normal atmospheric conditions of about 18,000 btu per pound, and, because it is a room-temperature liquid, one cubic foot (or about eight gallons) has almost one-million btu of convenient energy at atmospheric pressure. The container doesn't even need a super-tight cover. Natural gas (primarily as methane with one carbon per molecule) has only one thousand btu of energy per cubic foot at atmospheric pressure of 14.7 psi. At atmospheric conditions, almost six-thousand cubic feet of natural gas are required to equal the energy in one 42-gallon barrel of liquid hydrocarbon fuel, like gasoline or diesel. If the natural gas is compressed at 3000 psi, or super-cooled to a liquid at –160°C (both require great energy input and extremely complex and dangerous storage), the volumetric energy density of natural gas is increased from 1,000 to about 200,000 btu per cubic foot, **still only about one-fifth the energy density of liquid petroleum fuels.**

Low energy density is why natural gas is called a "stranded" resource. It is limited to low-pressure overland pipeline distribution, or must be cryogenically cooled to a liquified natural gas (LNG) with a very expensive and complex infrastructure for intercontinental shipment.

To convert and transport natural gas as a LNG requires (energy lost) about twenty-five percent of the energy in the gas. As with oil, there are politicians who promote the infrastructure and export of LNG to exploit the three-fold cost difference between surplus "fracked" gas in the U.S. and the price other countries are willing to pay.

Any talk of CNG at 3000 psi for motive power infers short distances, extreme danger, frequent filling with complex equipment, and very heavy, expensive fuel tanks. The only available CNG passenger vehicle is the Honda Civic which has a heavy, chrome-moly fuel tank that requires a substantial part of trunk space, and is rated at eight gallons of gasoline equivalent (GGE). For this inconvenience, the Honda customer is expected to pay a premium of $8,000. A quick Google search will provide copious further information and glowing publicity about CNG vehicles. Instead, try: Bernstein.com, see *"Why we don't see natural gas vehicles putting a dent in gasoline demand"* (1/4/13). Despite T. Boone Pickens' glowing prophesies, the time, investment, and energy required for a meaningful transition to a CNG transportation system are far too much to supercede the oil age. Would you take a plane trip on a high-pressure fuel tank with wings?

Another proposed alternative is to chemically convert the "stranded" surplus of natural gas to a liquid fuel called gas to liquid (GTL). Like coal-to-liquids (CTL) described below, the technology has been around for years but is extremely expensive and energy inefficient. **Recently, the Shell corporation spent nineteen billion dollars to build, by far, the world's largest GTL facility in natural gas-rich Qatar.** (Ref.: "Shell's Pearl proves its worth, but it's early days yet for gas-to-liquids," europeanenergyreview.edu/artikel3846). This city-sized operation has a capacity of 250,000 barrels per day, almost equal to all other world-wide GTL pilot plants, and proposals, combined, **and about three percent of present US gasoline consumption**. The investment costs, and intellectual secrets involved, are significant, certainly enough to discourage further enthusiasm.

Nevertheless, the recent excitement over "fracked" natural gas is "fueled" and typified by a August 15, 2012 article in the "Wall Street Journal," *The U.S. Natural Gas Boom Will Transform The World,* and subtitled: *North America's massive resources are going to shift market power away from OPEC and Russia and to consuming nations.* **Then, this feature immediately segues from natural gas to the North Dakota oil Bakken/Three Rivers oil boom presently producing about 600,000 barrels per day of oil, including highly-flammable condensate.** This "boom" presently amounts to about eleven percent of total U.S. oil production, about three percent of U.S. consumption, and less than one percent of world extraction as shown in Figures 3 and 7; hardly enough to "transform the world" or, as stated further in the above reference: "A United States hopelessly dependent on imported oil and natural gas is a thing of the past." Many energy experts now predict that North America will have the capacity to be a net exporter of oil and natural gas by the end of this decade. This is an example of gross disinformation from the media, and certainly enough to confuse and placate the public. Refer back to Figure 2 to see how many years our present lifestyle will last without import or export of oil.

For a more pessimistic example of recent discussions about "Energy Independence", visit the website alternativeinsight.com. In a Q&A session, the host, James Stafford (editor at oilprice.com) asks energy expert, Chris Martenson (author, *The Crash Course*): "Should the U.S. export natural gas?" Answer: "Fossil fuels. They're a one-time gift. You get to extract them and burn them exactly once. That is whatever you choose to do with them is what gets done. They perform work for us. So we really should be focused on what sort of work we want those fossil fuels to do for us. There are, right now, about a dozen proposals to liquify and export US natural gas, and a study just came out this past week, commissioned by the EIA, saying that's a good idea. Wrong, it's a terrible idea. Fully 25% or more of the energy contained within the natural gas is expended just in the process of liquefying it. That's what you get to do with 25% of the units of work. You get to turn the gas into a liquid and nothing else. We should be using every possible unit of work that we extract from the ground contained within that natural gas to do something useful. If it were mine to say, we'd be using that energy to rebuild our nation's crumbling infrastructure; we'd have a 30-year plan for exactly what we want our country to look like and how we're going to use our natural gas to get there. So when the natural gas runs out, and it will someday, we'll at least have a resilient well-built country that can run on alternative energy sources."

To this conversation, I would add: it certainly appears that greed (the "Selfish Gene"), as manifested by short-term profit motive, completely dominates public and political discourse with no consideration whatsoever for what's left for our children. This is exactly the same attitude that entices us to buy the things we want today and put the charges (plus interest) on our credit card. **This system only works when the continued growth of resources can support concurrent, continued economic growth to pay the bills plus interest to investor (banks) who really add nothing of substance.** On a longer term national level, we expect that continued energy-fueled economic growth will allow an increased population to pay back today's increased borrowing. Currently there are over a trillion dollars of student loans, almost a trillion dollars of credit card debt, and nine-hundred billion dollars of auto loans outstanding in a country that is eighteen trillion dollars in debt. Declining energy, beginning with oil does not support this traditional way of thinking. **Trivia fact: a trillion dollars is a seventy-mile-high stack of $1,000 bills.**

How much natural gas is left?

This is another of those difficult questions. When I first ventured into energy activism about ten years ago, the future of natural gas was even more suspect than for oil. There were sporadic shortages (remember Enron?) and wholesale market prices were pushing fifteen dollars per thousand cubic feet (with natural gas, each cubic foot has about one-thousand btu, so one-thousand cubic feet has one-million btu of energy, about the same as nine gallons of gasoline, or one-fifth the energy in a barrel of oil). The U.S. was already importing LNG through four ports. Julian Darley wrote the book, *"High Noon for Natural Gas."*

Then, horizontal hydraulic fracturing ("fracking") completely changed the picture. In the last five years, there has been much euphoria and the resultant glut has kept the price struggling to exceed three dollars per cubic foot. Keep in mind natural gas (like electricity) is very difficult to store and has to be used immediately when produced. There must be a pipeline (like an electrical transmission line) directly connecting the source and user.

With that background, Figure 6 shows the *Maggio* contemporary reference source for natural gas in the same units as oil and coal. Peak world extraction is projected to peak in 2040 at an annual, oil-equivalent rate of twenty-three billion barrels a year. This predicted peak is later than oil but ten years earlier than coal. Any attempt to more accurately predict and ensure the future of natural gas requires a higher price to offset the short life of "fracked" wells where the extraction rate can quickly drop to one-half after the first year. For a recent comprehensive analysis: enews.net/energywire/2013/02/11. **But remember again, (and again!) the peak for all three fossil fuels is well short of one-half the lifetime of a child born today.** At least natural gas is the cleanest fuel and will have less environmental impact. More on that subject as we get into coal.

KING COAL

As with natural gas, our modern-day economy and lifestyle cannot possibly continue, as-is, by shifting energy dependency from oil to coal. True, coal can and has been used as a transportation fuel. In the mid-nineteenth century, easy-mobility as a key feature of the industrial age, began with coal. It was carried on-board and used at very low efficiency of around ten percent, along with lots of water, for high-pressure steam motive power in ships and locomotives. Modern-day consumption of coal is primarily in power plants for the production of electricity, about seventy-percent of China's total energy, and twenty-five percent in the U.S. Our contemporary electrical system can continue beyond the near-future demise of the oil age by using a mix of natural gas, coal, nuclear and hydro. But all except hydro are finite fuels and none can support transportation and agricultural energy requirements without batteries, power wires, or third rails for mass transit. As with natural gas, there is no place for airplanes in this discussion.

Also, coal has always been fundamental to the industrial age as the primary energy input for smelting ore or scrap into every conceivable size and shape of cast iron, steel and other metals. Think of every giant skyscraper as a testament to cheap energy. All modern cities built their skyline, interconnecting rail, and ocean-going infrastructure in the age of coal and steel. But each city also owes just as much to electricity, natural gas, and oil to keep it running.

The technology for converting coal to a liquid fuel (CTL) has been around for years as exemplified by Hitler to fuel his WWII efforts as an alternative to Germany's meager petroleum endowment. Currently, there are still small CTL pilot plants around the world, especially in China; but the conversion efficiency is only about fifty-percent and the pollution is terrible. The CTL alternative could not possibly outweigh the hurdles of time, capital, and energy input required to supercede the oil age. Besides coal, itself, is a finite fuel and extensive conversion to a transportation fuel, including electricity, would only hasten depletion.

How much coal is left?

The optimists infer we have "hundreds of years left." They simply divide seemingly infinite reserves by the present consumption rate. To better answer the question, I will defer to two reports: the world-respected Energy Watch Group (EWG) comprehensive report (final version28032007), and the G. Maggio report: "When will oil natural gas, and coal peak?" shown in Figure 6. Both these reports are in good agreement and show vast reserves of coal with an average oil energy-equivalence of one ton of coal equaling about three and one-half barrels of oil.

About twenty-five percent of the world's coal has been mined to date. The EWG report projects a similar near-term increase in production as Maggio, but with an earlier peak in 2025 due to environmental concerns. **Neither report discusses how much of the remaining reserves can be mined and shipped without oil-energy input as diesel fuel to do the work. Both reports predict a peak and contraction of the coal age in less than the lifetime of a child born today.**

Another contemporary, peer-reviewed paper by T. Patzek and G. Croft is titled by the controversial statement: *World's Peak Coal Moment Has Arrived* (nytimes.com /70121). Clearly, there are hundreds of billions of tons of coal left; and the world, led by China at four-billion tons per year, is presently mining and consuming about seven billion tons per year. This is the equivalent energy of about 24 billion barrels of oil assuming median energy-quality for the different grades of coal.

POLLUTION, CLIMATE CHANGE, AND GLOBAL WARMING

One thing is certain in the coal discussion. Because of its high carbon content **plus all the other pollutants like lead, mercury, sulfur, and arsenic,** a post-oil age dependent on increasing energy proportion from coal will not be a healthy place.

We are quickly moving toward a run-away, 450 ppm CO_2 climate, well past the idealistic limit of 350 ppm. **Clearly, after the end of our two-lifetime oil party, the environmental hazards of attempting a coal-age sequel will most-certainly change life as we knew it.** Any talk of carbon sequestration is just that ... talk. It's not being done, anywhere. The Chinese are building a new coal-fired plant every week and donning gas masks in Beijing. It's difficult to get excited about any type of "going green" efforts for a sustainable future when coal-fired utilities totally dominate the steady deterioration of our environment. There are myriad new books that delve deeply into environmental issues. One broad summary of related essays is, *Fleeing Vesuvius, Overcoming the Risks of Economic and Environmental Collapse* (new society Publishers, 2011). Unfortunately, even in this book, the ever-present backdrop of continued population growth is absent.

As a footnote, Britain was totally dependent on coal at the turn of the twentieth century just after Stanley Jevons correctly warned that increased efficiency would only lead to even greater consumption and hasten depletion (Jevon's Paradox). This is when the steam engine-powered pump ushered in the industrial age and allowed miners to dig deeper for the remaining coal. Fortunately, the oil age rescued England from "peak coal" and terrible London smog just as the UK supply peaked at about the time of WWI.

Figure 6 also shows the EIA predicted growth of CO_2 emissions that are coincident with their projected fossil fuel increases through 2035. **The world's total CO_2 projected increase is forty percent from 31,305 million metric tons in 2010 to 43,220 million metric tons in 2035.** There is an extremely vociferous counter movement that still argues that greenhouse gasses are not a problem, nor is global warming a modern phenomenon or anthropogenic (caused by man). I can't understand why so many want to debunk what is claimed to be the most serious problem facing humankind. If you want to jump into the middle of this cat-fight, just contrast www.350.org with www.icecap.us, or www.wattsupwiththat.com. If that's not enough, listen to the Rob Hopkins interview with Michael Mann (*The Hockey Stick Wars,* www.transitionculture.org).

The EIA and DOE have a very disturbing pattern of predicting growth of everything, including CO_2 emissions, until 2035 and then leaving us hanging with no concern whatsoever of "what next." Spend some time on eia.gov until you are buried in statistics, projections, and dissimilar units of energy with seemingly no concern for the overall concept of "finite" resources or the future after 2035. See additional EIA discussion on p. 51, "EIA Dreamland." On the contrary, the DOD is very concerned about the future of its navy in light of peak oil, climate change, an open arctic, and rising ocean level. Read: "Full Green Ahead," *Mother Jones*, March-April, 2013.

Another recent, very disturbing and directly-related (to climate change) report can be found on AMEG.me. This analysis of the rapid decline of arctic sea ice is the work of the Arctic Methane Emergency Group and was just released in December, 2012. To quote the group's chairman, John Nissen, "It's all about the arctic sea ice … Abrupt climate change is upon us … Food prices will go through the roof … complete summer collapse is expected by 2015 …putting us in a state of planetary emergency today." This is another example of ubiquitous environmentalism. If those liberal, "enviros," "tree-hugging-greenies," whatever; would just give free-market capitalism a free rein, all will be ok. We can live life to the fullest today including continued growth and prosperity with no consideration for the quadruple crises: energy, population, ecological devastation, and economic collapse staring us in the face. See: robbwiller.org, *"The Moral Roots of Environmental Attitudes."* I have long been of the opinion that energy, specifically peak oil, will be the most urgent and critical of the four, but obviously, all are inextricably combined and must be considered together, **especially since 2012 was the hottest year on record until superceded by 2014. Again, as admonished throughout my humble effort, nothing at all will happen unless you help make it happen! Please get involved. Reread Chapter 3.** Back to the energy update:

FINITE NUCLEAR

Uranium, as used for nuclear fission, provides about thirteen percent of world electricity, and six percent of total energy, but it is also a finite fuel. A quick "Google-U" search shows a third peak of extraction rate at 120,000 metric tons (2200 pounds) expected about 2050. This compares to a current consumption rate of 70,000 metric tons with fifty-percent of the ore coming from Australia and Kazakhstan. There seems to be plenty left, and there are many optimistic theories for improved reactors, alternative radioactive fuels, and "unlimited" reserves at lower concentrations. There is a wide disparity in use of nuclear power among nations. France relies on nuclear for seventy-eight percent of its electricity, the U.S. only twenty-percent. The total contribution of nuclear power for energy is shown in line "N" Figure 6 in terms of equivalent billions of barrels of oil per year. In addition, the catastrophic accidents of Chernobyl and Fukushima along with the threat of terrorism and monumental start-up costs have soured public acceptance for nuclear as a major electrical-energy source. A recent study predicts unprecedented disaster for all of western Europe if there is a nuclear plant blow-up in France (see: www.eurosafe-forum.org).

All of the above considered, it is doubtful if nuclear will ever exceed its present world-wide energy contribution of six percent. But, like natural gas, nuclear, including liquid fluoride thorium reactors (LFTR), and hydro could help as a bridge, "base-load" energy source in a sporadic wind and solar-electric future. This is significant because it will take a long time and considerable capital for an energy-transition to wean ourselves off fossil fuels, cope with climate change; and most importantly, reverse population growth with a proportionate need for food.

EIA DREAMLAND

Also shown in Figure 6 are recent energy forecasts by the U.S. Energy Information Administration (EIA) World energy consumption profile (Wikipedia.org/world-energy-consumption-outlook, August, 2012, and EIA.gov/international energy outlook). The four projections show continued linear growth, from 2012 to 2035, for all four finite-fuel energy sources (including "yet to be discovered").

This same statistical wing of the U.S. DOE predicted flat-lined, short-term retail prices for all three fossil fuels plus nuclear electricity. For instance, retail gasoline was projected in 2012 to drift downward to $3.34/gallon by July, 2014. Now, in hindsight, we see that gasoline climbed back to the $4.00 gallon range in February 2013, and then plummeted to below $2.00 by January 2015. Apparently, the EIA thinking is that steadily increasing oil availability and continued world recession will combine to stabilize prices even in the face of growing population and high-energy lifestyles. In the commodity markets, basic underlying movements of supply and demand can be rapidly magnified by speculation and the physical limits of storage of the commodity itself.

To reemphasize a basic theme of this book, and as shown in Figure 1, the American motoring public is, by far, the dominant customer bloc for oil. As Americans steadily, and inequitably, go broke, the market price for oil and gasoline is dragged down ("demand destruction") to the delicate balance-overlap (if any) between declining, remaining wealth available to buy, and the diminishing number of suppliers (including U.S. and non-conventional) ready and able to produce at the reduced market price level.

CONCLUSION

To end this brief overview of finite energies, one thing remains evident: All finite sources will surely peak, and significantly decline, in the next one-half lifetime of a child born today. Renewables cannot possibly climb to higher than

ten percent of oil alone and certainly not to ten-percent of all today's energy. What then? It appears certain, and typical of human nature, that short-term greed will trump long-term planning. **Oil is the most important fuel as it is fundamental for ninety percent of modern transportation, nearly all of agriculture, and the support of the production of all other energy sources.** Meanwhile, population just keeps on growing. The reversion to dependency on coal as a critical pillar of our future is unnerving. Emissions and climate change will continue to increase, meaning even less food for increasing numbers. As with past failed societies, weather will be blamed for the crash rather than population which always increased when times were good. The cleaner energy sources like natural gas and nuclear can help maintain an electrical-powered world, but neither can offset the demise of oil for travel, construction, transportation, and large-scale-agriculture (see Appendix A).

Also, it is important to remember that the end of the fossil fuel age will also be the end of thousands of other products we take for granted, use daily, and are completely dependent on for hydrocarbon feed stocks. The end of the age of travel will also be the end of the age of plastic. There is nothing we can do about this. A tiny bit of bio fuels, bio lubricants, and bio plastics will only compete with precious food, especially without inexpensive energy input from oil. This composite of dire reality could be partially, and immediately, mitigated by equitable coupon-rationing of gasoline, our most-wasted fuel. Does Chapter 7 now make more sense?

RENEWABLE ENERGY SOURCES

Beyond the ephemeral age of finite energy sources, which presently supply over ninety-percent of the world's total energy, renewables, specifically solar, wind, hydropower and biofuels are our only hope for a future with any semblance to the lifestyle we take for granted today. **Finite sources** (except for nuclear) represent stored energy from eons of ancient sunlight. **The renewable sources are severely limited because, instead of millions of years of accumulation and convenient storage, they are dependent on current dilute and sporadic solar energy.** All renewable energy sources are shown as line "R" in Figure 6. Other sources of energy like tidal flow from the gravitation of the moon, and geo-thermal heat from deep in the earth, although renewable, are so limited as to be non-issues. Before the industrial (fossil fuel) age, renewable energies supported all living species. Growth was limited by current solar input and minimal storage capacity like a dam for hydro power or a tree for bio fuels.

Hydro power

Thermal energy from the sun transports water vapor from the oceans to higher, colder regions. There it condenses, collects, and flows downhill thus converting potential energy collected and stored in dams, into kinetic energy to turn water wheels or turbines. Simple enough, but limited by the height, area, and topography of land surface required for collection before it flows back to the oceans or evaporates on route. Except for still-undeveloped potential in a few remaining mountainous areas, these constraints plus massive capital investment and regional disruption like the Three Gorges dam in China, limit hydro power to its present contribution of about seven percent of the world's energy. Also, it is only a source of electricity. It's an excellent electrical energy source for mass transportation only in developed mountainous regions like the Alps or Scandinavia.

Hydro power, as potential energy stored in a dam, resolves the energy storage problem and is excellent for backing-up intermittent wind or solar. Pumped storage uses temporary excess power to pump water uphill to a storage reservoir. This stored energy could also be used as needed to meet intermittent demand. Pumped storage can be very efficient and, on a regional basis, is one of the best ways to balance varying inputs and demands.

But, water flow is also impacted by climate change, which has further reduced hydro's contribution to America's energy mix in the last decades. Back in the 1930s, hydropower was the largest source of electricity in the U.S. Now it is a minor player at about fifteen percent of electricity and has been relatively flat since the mid-1990s because the best sites are in use, and water flow is declining.

Renewable bio fuels, wind, and solar

As shown in the table at the beginning of this chapter and Figure 6, **all renewable fuels minus hydro contribute not more than three and one-half percent of total world energy**. This is the equivalent energy of three billion barrels of oil per year. As shown at the very bottom of Figure 6, wind and solar (W+S) together contribute less than one-percent of the world's energy today. This leaves less than three-percent from renewable bio fuels some of which, like ethanol and bio diesel, are also included as a small percentage of all-liquid fuels. Also included in the "everything else" category are laboratory or pilot-plant scale (and very poor EROEI) energy sources like algae, tidal power, and cellulosic ethanol, which do nothing but mislead and divert public attention from the imminent crisis of finite oil.

Traditional biofuels, like wood, would not even come close to keeping today's population warm, especially with diminishing liquid fuel energy-input for harvest and delivery. Also, dependency on biofuels in any form must always respect the regrowth time for sustainable harvest and return of nutrients to their source. Otherwise, it's Easter Island all over again. **All non-renewable energy sources together, including hydro power, can never exceed more than ten percent of today's world energy consumption.** They are only different forms of daily solar energy with the same limitations of dilute sporadic solar input, poor EROEI, and difficult storage.

But, we might ask, how did colonial Americans survive and proliferate with only an axe and a horse or ox? They even had time to clear the forests, have ten kids, and build the stone walls like those that surround our beautiful old farm in Maine. (The "three-holer" in the barn, hand water pump, horse-powered hay-wagon unloader, and spring water system are still here.) The best answer for their success is that early farmers, although quickly increasing in numbers, were still well below the carrying capacity of their prolific, immediate surroundings. The sparsely-settled country was teeming with fish and game. **Very hard work cleared the fertile land. Farms were built with two and four-legged muscle power.** Life was hard and short (for many). It is my dream that we, as a nation, could gradually return to a similar sustainable life style along with a stabilized, dependable climate; but with far fewer people than we have today. Adding to my limited optimism is the opportunity we have to benefit from a modern electrified lifestyle as described in Chapter 5.

NON-ENERGY, NON-RENEWABLE RESOURCES

Starting back in the iron and copper ages and progressing forward from spear heads and plows to windmills and guns, **it is obvious that the progress of humanity was completely dependent on finite metal resources.** With only wood and charcoal, and no fossil fuels for mining, smelting, and fabrication, there probably would have been enough metal ores to last for thousands of years. Now, however, every non-fuel, under-the-ground, finite resource is being rapidly depleted and the reserves of some are more critical than oil. **It may well be that no advanced civilization can exist very long on a finite planet because all finite (by definition) non-renewable resources will eventually be depleted.** Exactly like finite fuels, the competition between growing population and diminishing non-fuel finite resources is first economically defined (price), and then usually becomes belligerent as scarcity separates the wealthy from the masses. The best book about all non-renewable resources is Chris Clugsten's *Scarcity, Humanity's Final Chapter* (see Chapter 11 for review).

PART III

DOWNSIZING U.S. OIL CONSUMPTION FROM 22 TO 3 BARRELS PER PERSON PER YEAR

The preceding chapters defined the impending world energy crisis. A closer focus shows quantatively that the egregious U.S. oil consumption of 22 b/p/y is the dominant singular factor. This inordinate share cannot continue more than several more decades as explained in various scenarios in Chapter 2.

Climate change is directly related to oil consumption and is the subject of much political and media argument and obfuscation. Unfortunately it completely diverts attention from the far more serious and imminent demise of the high-energy American way of life.

The next part includes in-depth exploration of specific related subjects that by themselves have direct bearing on, or offer prospects for drastically downsizing our oil addiction, as we must do immediately.

CHAPTER 5

A Solar-Electric Future

OVERVIEW

The only concievable path that leads to, at best, a tiny fraction of today's finite-fueled, high-energy age is to move as quickly as possible into a sustainable future powered primarily by hydro-electric and variations of incoming solar and wind power. A growing solar-electric future could overlap and extend all present sources of primary electricity. Unlike hydro, PV panels and wind turbines are scalable from their present tiny contribution of less than one percent of the world's energy; but only if we move quickly, and as long as there is still surplus oil and remaining wealth left to "fuel" the transition. **We must also anticipate and respect the extreme technical complexity, capital investments, and non-energy material requirements required to vastly expand and perpetuate a solar-electric age.**

There are many troublesome obstacles to this future direction, such as infrastructor and the limitations of electrical energy storage and motive power. For example, the book *Green Illusions* starts right out with Chapter 1: *Solar Cells and other Fairy Tales*. Another voice is the political sarcasm in *The Solar Fraud, Why Solar Energy Won't Run the World* (Hayden, Vale Lake Publishing, 2004). Hayden has a Ph.D. in physics and completely understands the acute, quantitative limitations of attempting to scale-up and substitute contemporary solar energy for hundreds of millions of years of conveniently stored ancient sunlight. Both books are correct in delineating the limitations of renewable solar, especially cost and sporadic feebleness. They are quick to point out that solar power cannot remotely approach the present energy levels of fossil fuels. **But, these nay-sayers do not offer any alternatives, except dig more and drill deeper, nor do they venture into directly related discussions about food and population.**

Today, about 90% of the energy we use for travel comes from oil. We will never again go far or fast without it except within the narrow limitations of wind, coal, wood, and electricity. Winter travel will be especially problematic and we will never fly without gasoline or jet fuel. Liquid bio fuels are not the answer because of their poor energy return (EROEI) and use of crop land instead of for food. The age of air travel began just over one lifetime ago when the Wright Brothers first added a gasoline engine to a glider. Lead acid (L/A) batteries that weigh 80 times as much (20 times as much for Lithium-ion) for the same energy content as petroleum fuels, cannot lift their own weight off the ground at speeds sufficient for a wing to provide lift. Unless pressurized to 3000 psi in a heavy container, natural gas also does not have the energy density to power a plane, vehicle, or boat. Besides, the extreme flammability would be unacceptable. (Remember the Hindenburg disaster.) **As stated throughout, and is the underlying theme of this book, our only hope for transition to a tiny fraction of the short oil age is to carefully nurture (ration) our remaining oil starting immediately.**

PERSONAL EXPERIENCE

Over a decade ago when I first became convinced and alarmed about the steady depletion of fossil fuels, I began building and testing solar-electric concept vehicles to explore and define a future without oil.

The magic, convenience, and dependability of a PV cell (when the sun shines) never ceases to amaze. However, I would summarize my extensive experience with electric vehicles by saying: without grid back-up from other energy sources, **any form of solar power is too feeble to directly replace even a tiny fraction of our present usage of liquid fuels especially for motive power.**

The first tractor and MG

A small 1954 ten horsepower Farmall Cub tractor with a permanent on-board 750 watt (one horsepower) array is shown below directly charging a battery-powered

lead/acid (L/A) 1962 MG. This tractor has worked dependably for us since 2005. It has been to many fairs and was featured in *Mother Earth News* (August 2006). **Its primary shortcoming is feeble energy storage of 5 kWh, equal in work output to one-half gallon of gasoline, from nine Walmart deep-cycle batteries. Also there is the need for maintaining the batteries and recycling them every few years (more on that later).**

Next, I restored the MG midget in 2006 with a high current, 150 ampere hour, 72 volt battery pack. At just over 45 amperes (about 5 hp) at 30 mph, and with antique registration plates, I was able to drive the car over 100 miles during each of several controlled tests. However, the 800 pound battery pack was too heavy for hills and the car's tiny brakes. I then changed the battery pack to ten Walmart "MAXX29" batteries which reduced total battery weight to 600 pounds in a stripped-down 1200 pound car. This 120 volt system voltage reduced the amperage (proportional to power) to as low as 30 amps at 30 mph (one ampere hour per mile), but the lower battery energy capacity also reduced the usable range to about sixty miles. At least the car was "zippier" and would go up hills at 30 mph as long as the current did not exceed 100 amperes for more than a couple of minutes; otherwise, the battery energy capacity and range would be further inordinately reduced. (Google: Peukert's exponent.) The MG was a lot of fun on back roads, but I don't register and insure it any more because it is definitely not compatible with the American propensity to drive huge vehicles, much faster than necessary, to urgently get somewhere else from where they were. (See Chapters 1 and 7.) Also, the construction, brakes, and altered power system do not meet Maine inspection requirements for antique vehicles.

Scaling up motive power

A larger 1500 watt stationary solar array equal to two horsepower, and twice the size shown in the photo on the Cub, could collect up to nine kilowatt hours of energy in a typical long summer day. That much energy could do the same mechanical work as one gallon of liquid fossil fuel in an internal combustion engine at 25% efficiency.

An equal amount of human labor would require one hundred and twenty hours of tedious human work at an average power output of seventy-five watts. This much work is the same as **fifteen, eight-hour days (three 40 hour weeks) of labor including the simultaneous human need for 120,000 btu's (35 kWh) of food-energy input. This much energy for nine kilowatt hours of output represents a deficit EROEI of 0.25:1; not a very good proposition compared to PV cells or conserving (rationing) remaining liquid fossil fuels.**

Home (distributed) power

Eight years ago, in my book, *The End of Fossil Energy,* **I proposed 40 million homes (about the number of private residences in the U.S.) each having a 4,000 watt (peak PV) personal decentralized (distributed) array ... in 40 years.** Each system would be, in essence, a private free-enterprise electric utility with great efficiency and complete freedom from oil shortfalls, power failures, distribution losses, terrorism, and utility profits. With this system, the average home in the northern latitudes of the U.S. could expect to harvest an average of 400 kWh per month (4800 kWh/year) at a 14% capacity factor (sunshine hours per total hours). **With this 400 kWh per month "energy budget" the following needs could be supplied:**

- 200 kWh for personal transportation in a battery electric vehicle (bev) similar to the MG. This would be **enough for up to 1600 miles of travel, or 55 miles per day at one ampere hour per mile.**
- Instead, this same 200 kWh could be used, during farming season, to provide up to 20 hours of tractor power (at 10 kw or 14 hp) in one of the larger Ford tractors described next. **This is enough serious power and energy to plow and harrow about 5 acres of land for planting.**

The remaining 200 kWh could energize:

- 85 kWh for hot water heating (preferably augmented by thermal-solar), enough for one shower or bath every two days at 3412 btu per kWh.
- 15 kWh for full-time of a high-efficiency, full-size (12 ft^3) refrigerator.
- 45 kWh for 30 hours (1 hr/day) of cooking on a hot plate
- 12 kWh for 4 hours per day using the TV, home entertainment, and computer.
- 8 kWh about 80 hours with the washing machine.
- 12 kWh 720 hours of lights. (Two thousand hours with LEDs)
- 12 kWh typical water pump.
- 11 kWh miscellaneous power tools.

Total: 400 kWh for everything including minimal transportation and agriculture.

When I wrote the third edition of my book nine years ago, installed PV systems cost much more than now. Chinese competition and economy of scale have forced our domestic PV production to be much more efficient. Today, at a panel cost approaching $1 per watt, instead of $5 per watt 10 years ago, a complete 4 kw system would cost about $6,000 including labor and grid-tied interface, **but no independent battery back-up.** My total plan of 4 kw for forty million homes, in one year, would supply 0.655 quadrillion btu to our nation's energy, and **displace 0.3 equivalent billion barrels of oil** or other fossil fuels burned at 33% efficiency, all with practically zero pollution.

To put this much renewable energy into perspective with today's profligate fossil fuel consumption, 0.3 billion barrels of oil is less than one tenth of our annual domestic gasoline consumption or 1% of world oil consumption. We could save this much energy just by slowing down 10 mph! (See Chapter 7.) It's astonishing to me that we do not have the collective willpower to understand these facts and get going as time is quickly running out.

Solar-powered golf cart

My third solar-electric vehicle, the golf cart shown in the next photo, is our favorite. It's a standard 48 volt commercial Club Car with its own 350 watt, two-panel onboard array, and a 2500 watt (3.3 hp), 120 volt ac inverter permanently installed under the small pick-up bed in the rear. With four 1.5 kWh batteries, it can travel 80 miles at 15 mph, **without any solar input**. There is no need for range anxiety or tired batteries because you can always stop and recharge, as long as there is sunlight.

The golf cart requires about 24 amperes (1200 watts or 1.6 hp) of power for average travel as long as there is minimal braking (energy lost as heat) instead of coasting. It is especially convenient on our farm as a mobile power source for any 120 volt tools. For instance, as shown, the on-board inverter is perfectly adequate for a 3 ½ hp electric chain saw. Anybody who questions the efficacy of solar power should try cutting firewood the old fashioned way with a cross cut or a buck saw. The golf cart is also used as a back-up power source for our home when the utility power fails.

Larger tractors

Two additional tractors were built next in 2007 and 2008. A 1948 Ford 8N and a nearly eighty-year old 1938 Ford 9N, were also converted to test electric power with full-size 20-plus horsepower tractors. These tractors have modern Category I, 3-point hydraulic implement hitches first introduced in the Ford 9N. The rear power take off (PTO) provides power for any number of modern attachments. The energy storage is in ten 120 pound, 12 volt L/A industrial batteries connected in series. When new, the total 1200 pound battery pack, fully charged at up to 140 volts from a grid battery charger or PV array, can store the equivalent energy at 75% depth of discharge (DOD), and current draw less than 75 amperes (12 hp), to do the equivalent work as 1½ gallons of gasoline (11 pounds). This is enough for about one hour of serious plowing (16 inch single bottom) or harrowing (six-foot double disc) at up to 20 hp. But it would take at least 16 hours of direct sunlight shining on a 1 hp on-board array, as on the Cub, to recharge one of these 12 kWh battery packs. The 8N is shown below with a full 16 inch single-bottom plow while harrowing simultaneously.

The 1939 9N in the last photo is powering a PTO-drivin, 5-foot rotary mower (bush hog). Power requirement for this task, on the level and low gear, is 75 amps times 120 volts or 12 hp. Large electric motors need to run at higher rpm than normal 1500 rpm tractor engines. The 8N utilized a low ratio Sherman transmission and the 9N uses a 1.6:1 tooth belt drive to keep the electric motors spinning faster and cooler.

These two tractors have been performing admirably for years but need frequent rests for motor or controller over-heating. They are extremely dangerous because of high voltage DC and lack of compression braking as with a familiar gasoline or diesel tractor.

This solar-powered agricultural exercise drives home the reality that the high-power commercial machines we take for granted like 18-wheelers or earth movers will never work in a low-power solar-energy age ... all the more reason to ration the oil we have left and not waste it on frivolous trips or entertainment.

Conclusion

To emphasize a rule of thumb worth remembering: **a 1500 watt PV array in one long day of summer sunshine will provide the energy of a gallon of gas, about nine kWh.** This is equal to twelve horsepower-hours of energy to do meaningful work. This may not sound like much, but it will power a small electric vehicle like the MG for sixty miles or an electric tractor could plow or harrow one-quarter acre (11,000 square feet) of farmland in an hour. If not needed for transportation or farming, the same energy could supply the electrical needs of a typical, energy efficient home (using 100 kWh per month) for three days. **This much serious energy would easily resolve the problem of "powering the farm" without fossil fuels or draft animals. One farmer could conceivably farm three or more acres and therefore feed twelve or fifteen non-farmers.** He and his family could also enjoy a "modern" lifestyle without physical exhaustion and go for a ride in a battery-electric-vehicle (bev) once in a while and/or deliver food to villagers. **It could be done ... except:**

ENERGY STORAGE AND BATTERY RECYCLING, the ACHILLES HEEL

As with all forms of energy, storage is the critical weakness. You cannot borrow new incoming energy from the future for use today. This is why a debt-based financial system is illegitimate for the future purchase of energy in an age where energy is declining. Except for direct solar heating, which provides warmth only when the sun is shining or for a few additional hours from an adjacent warm mass, the needs of all humans are far greater than can be supplied on a 1:1 ratio from the weak, sporadic power from direct solar radiation. Plants address the energy storage problem with photosynthesis. This is the chemical process of using incoming radiant energy to combine atmospheric carbon dioxide with water to form and store high-energy complex carbohydrates.

The increased molecular weight of carbohydrates makes these compounds solid and stable (for a while) at normal ambient conditions and therefore possible to concentrate and store substantial quantities of energy for future use. A subsequent exothermic chemical reaction with oxygen (burning or metabolizing) releases the stored energy much faster (higher power) than it was slowly accumulated as solar input. The energy can then be used when we need it to keep us warm, fed, moving around, procreating, and all the other wonderful and bad things we do as intelligent primates.

The preferred method of electrical energy storage in the industrial age is the ubiquitous battery which uses a chemical process to store and supply electricity as needed.

Lead-acid batteries

The rechargeable lead-acid (L/A) battery has been the work horse for a century. If treated well and not deeply discharged, it can have a useful life of up to ten years. On the debit side, it is heavy, environmentally hazardous, inefficient, temperature sensitive, slow to recharge and quickly degraded if discharged quickly. But, there are millions of tons of lead in the world and lead can be recycled.

Our refusal to understand and tackle the lack of local recycling for L/A batteries may well be the one and only reason it will not be possible to segue from the oil age to a much lower energy solar electric future. It's that simple! Following that disturbing line of reasoning, I wrote directly to many who should be equally concerned, including the magazine *Home Power*, American Society of Solar engineers, Battery Council International, and numerous battery companies. **None ever responded.** Will anybody out there help address this critical subject?

In the first editions of my book *The End of Fossil Energy*, I showed, quantitatively, that **one 4 kw PV array (as on one of the suggested 40 million homes) could supply enough energy in one year to provide the heat of fusion to smelt 264,000 pounds of lead, enough for 4,000 batteries in a local community.** This is not rocket science. L/A battery recycling has been around for a century. If we were really smart, we would start immediately to simplify the L/A battery into several standard 12 volt designs that could be easily rebuilt (recycled locally) with maximum safety. Perhaps the container (plastic or wood) could be reused and the plates easily cleaned to salvage the sulfur. You can search web sites discussing this understandably-hazardous subject, even for do-it-yourselfers.

Electricity is not an energy source, but it is unique for instantly conveying energy, or for converting energy from one form to another like solar-electric input to mechanical output or heat. But, even more so than with natural gas or hydrogen which have minimal energy content and vary limited capacity for storage, **electrical power needs to be used exactly at the time it is produced unless converted and stored as another form of energy like the batteries discussed above, a dam of water, or a mechanical flywheel.** An exception is a capacitor which can accumulate electrons, but only at a very low energy level. If we are to rely on PV-powered homes or wind-powered utilities, we must focus on storing and smoothing the sporadic, intermittent incoming power and make it available as a continuous supply of dependable energy at the desired power level. Hydro electric power (and pumped storage) have done this for years. But hydro is only another form of limited and dilute solar energy collected from a very broad higher elevation land area, conveniently concentrated behind a dam. It can then be released as needed to flow downhill through a turbine and convert kinetic energy to electricity.

Residential storage

As we plan our idealistic solar-powered homes (for those few who can afford them) the storage problem is conveniently circumvented with grid-tied or fossil-fueled generator backup. For those few purists who are off the grid and no longer wish to or cannot access generator fuel, the only alternative is the same venerable battery, which inefficiently converts electrical energy to chemical energy and back again as needed. Chemical batteries are very inefficient if cold or not serviced regularly. In the future, after the fossil-fuel and nuclear ages, we will have to rely entirely on incoming solar (as PV, thermal, wind, hydro, and even biofuels) for all our electrical energy. Other than pumped hydro-storage or as energy stored in wood, giant flywheels, or caverns of hydrogen gas (the leakiest most reactive element of all) only the ubiquitous chemical battery can do the job. Considering the time-frame available, wealth to scale-up, and material tonnage (raw or recycled) required, the L/A battery is the still the best realistic candidate. The cost, raw material access, and transition time to lithium would limit that higher energy-density alternative to specialized applications as discussed later.

More battery concerns

Because of economics of scale and environmental hazards, all L/A battery recycling is concentrated in a handful (less than a half-dozen?) huge recycling facilities like the one operated by East Penn Manufacturing Company in Lyon Station, PA. **This arrangement works fine as long as the system is supported by a backbone of low-cost fossil-energy.** All we have to do is jump in our car or truck, trade-in the batteries at our local Walmart or specialized battery supplier, and we're good to go. Meanwhile, diesel-powered trucks accumulate the trade-ins, truck them thousands of miles to fossil fuel-powered recycling plants, and return with replacements.

Now, consider a very low-energy future, ultimately without any fossil fuels, but totally dependent on localized subsistence. Very soon our PV and battery-powered home, car, or tractor will need new batteries. Our tired battery-powered car or truck can't make it to a local supplier (if there is one) because the on-board batteries can no longer store enough energy from our personal, distributed PV system. We would have to call the local source and request delivery of the correct size. But the energy for delivery would also have to be battery powered. Remember, motive power and energy storage without fossil fuels are nearly impossible challenges. The L/A battery replacement function could conceivably be done locally only if it is carefully thought out and planned long in advance. **An absolute essential for every community center will be a L/A recycling facility, itself powered by local solar, wind or hydro power. The transportation of heavy batteries can not be more than about 30 miles, one-half the range of a battery-powered truck** ... unless, we carefully nurtured biofuels or hoarded (rationed) remaining fossil fuels specifically to support a solar-electric society.

To repeat, all L/A batteries are picked up and shipped long distance to one of a handful of huge, environmentally-friendly recycling centers. New batteries are returned by the same diesel-powered 18 wheelers. How can this be done without liquid fuels? Battery-powered trucks? And worse, like other toxic processes, we ship a substantial part of our battery recycling needs on to Mexico. According to a report from the NGO: Occupational Knowledge International, 261,000 pounds of used batteries (12% of all used batteries) and other lead scrap were shipped to Mexico to avoid the stringent environmental regulations in the U.S.

Lithium batteries

No discussion of electro-chemical energy storage would be complete without referring to lithium, the conventional wisdom for residential storage and personal transportation in the post-oil days. A recent book by Seth Fletcher, *Bottled Lightning,* is an excellent source for every aspect of the subject. I learned a great deal; for instance, there does not appear to be a shortage of Lithium in the world with vast deposits in Bolivia, Chile, China, and Nevada. Possible variations including li-sulfur, li-silicone, and li-air keep the dream going for a transportation future with lighter-than-gasoline energy and a 500 mile range. But remember: a lithium-powered future cannot possibly work without the time and immense capital investment for a transition. Besides, how can the lithium be mined and processed without ... oil? As my book is finalized for publication in 2016, lithium batteries have made the national headlines with a rash of fires in the just-introduced hoverboards. Charging and storage of energy in lithium batteries are notoriously dangerous.

WHERE DOES THE ENERGY COME FROM?

If we ignore or accept the electric vehicle limitations of range, cost, recharging infrastructure, and recycling; we must still come to grips with the energy-source problem. There is not nearly enough wind, solar, or hydro electric to meet even today's residential requirements without the "spinning reserve" of natural gas, coal, and nuclear. Where would the additional energy for transportation come from other than a small distributed personal PV system as described above?

How could the massive infrastructure for manufacturing and charging electric vehicles be funded? Would this investment come from oil companies, our government already in nineteen trillion dollars of debt, or consumers most of whom can hardly afford their next tank of gasoline? In what time-frame could an electric automobile system be built considering the imminent crash of our oil-powered civilization? How will highway maintenance be funded without a fuel tax? Why should a few wealthy buyers receive a rebate and, in effect be subsidized by all taxpayers, to purchase electric vehicles or install solar systems?

Complex battery management and fire hazard are additional major obstacles along the fantasy-road towards lithium. Or would we rather pin our hopes on Tesla and Elon Musk?

Other safety concerns

I have been asked for plans to build battery-electric cars or tractors. I won't help for two reasons other than not having the time or a commercially-acceptable product. First, anything above about 48 volts dc becomes **extremely dangerous.** A dc arc does not change polarity 60 times a second like ac. It just continues to arc over a longer gap and you can't let go! It will kill you instantly. In order to achieve adequate power for a car or tractor, ten 12 volt batteries in series are perfect. The total 120 volts dc brings the current down to, at most, 100 or 150 amperes, good for twelve kilowatts or sixteen horsepower. **My "concept" vehicles are very dangerous and I never have "hot" male terminals or plugs exposed.** Any home-built or commercial vehicles have to address these serious liability problems throughout the circuitry all the way back to a PV array or a 120 volt grid-charging, full-wave-bridge rectifier which I have built and use. Secondly, electric-powered vehicles do not have the compression braking we all expect from an internal-combustion engine. If you let up on the "gas," you just keep rolling. Perfect brakes, evenly applied to both rear wheels in a tractor, are critical. You can't put 'er in low gear going downhill with a big load of wood or hay and turn the key off and slow down with engine compression.

Normally, a charge controller is also used for PV charging, but I have never used one with my solar-charged vehicles because the battery pack storage is so large compared to the PV array input that overcharging is not an issue.

Solar-electric infrastructure

The next confrontation with reality in a solar-electric future is the complexity and availability of the myriad parts and materials that make it work. Everything from trace elements, to copper wire, to steel hardware and fasteners, to solid-state controllers, and on and on, are essential pieces of the whole high-tech system. All of these bits and pieces, **in addition to PV cells and panels, will have to be available without the support of an oil-based society.** This would be like going into a self-sufficient space-vehicle (the earth?) without a repair/support facility or parts. **We must conserve (ration) finite oil and prepare for a long overlap into a solar-electric age.** Many critical parts and materials should be stockpiled in advance. Non-energetic critical commodities will be needed and they, in-turn, are finite and need oil to mine, ship, and process as discussed in Chapter 4. Who will plan and pay for this nation-wide, forward-thinking leadership in our short-term, profit-motivated capitalism?

Wintertime blues

In the northern latitudes we also have to face the challenge of surviving the low-solar cold season and become less dependent on declining fossil fuels and utility-grid electricity but more dependant on stored food and biomass. Winter is when PV panels are most efficient. They produce more power at low temperatures because internal resistance is low. But, on the flip side, the days are short and frequently cloudy. **It is often a challenge just to harvest enough incoming sunlight to keep the L/A batteries from self-discharging, sulfating, and freezing.** There may be a sunny bright stretch of days with enough extra solar energy for a ride in a solar-powered car, but there will definitely not be surplus energy for occupant heating or vehicle lighting at night. Without oil (or a horse) we will stay put as in the "old days." At least we can go cross country skiing, enjoy our low-power LED lights, entertainment center, and modern communications, as we put another piece of wood in the stove and contemplate "one child per female" (1 cpf) as discussed next in Chapter 6.

Besides, how and why would the roads get plowed when oil distillate (diesel) becomes scarce and expensive? What types of vehicles will use the roads? **Considering the number of trips required by a typical diesel snow plow, it takes upwards of several gallons of fuel over the winter to clear each mile of road into four-foot snow banks. These solid piles of snow represent "stored energy" lost forever to melting in the spring.** Like heating and travel, this is just another form of wasted energy with no system-growth to show for it; all the more reason for gasoline rationing to restrict system-contraction as in the "energy barrel," (Figure 11 in Chapter 7). Several years ago I built a solar-powered snow-machine pulling a 48 volt PV sled-array. With four heavy, cold, L/A batteries, it did not have enough stored energy to push uphill through new snow without quickly discharging the batteries.

A "solar-slave"

Sometimes, with further thought, seemingly crazy ideas make considerable sense. **Consider the following "stimulus" proposal: The U.S. government will give, each year as a graduation present to every high school senior, a U.S.-made, 100 watt PV panel.** The approximately three-million panels at $100 each ($0.90 per watt plus shipping and economy of scale) would cost three hundred million dollars per year, barely any different than already lost in life-support for Solyndra.

This investment in future solar is about three-percent of our annual ten billion dollar budget to build fifty F-35 fighter jets. What will fuel these planes of the future, algae, switch-grass cellulosic ethanol? Think of a jump-started market flooded with PV:

- The next generation of job-seekers will have an introduction to future, post-oil energy.
- At an average, nationwide, capacity factor of 15%, the annual energy contribution from each panel would be 130 kWh (11 kWh per month). Referring back to the table at the beginning of this chapter, think what could be powered with that much energy: an electric bike, LED lights, communication, water pump, home refrigeration, all on a minimal off-grid "survival" basis.
- In effect, each recipient owns a personal slave, with equivalent power greater than one strong human, for upwards of forty years. There is no need for shelter, clothing, comfort, security, replacement, food-energy with dismal EROEI, and other human needs.
- A "street market" would immediately evolve for selling or buying panels as well as peripheral needs like switches, controllers, inverters, and electronic or analog meters.
- **The battery recycling problem would be given the serious attention it deserves.**

- At $0.15 per hour (today's electricity cost), one panel would offer an energy value of $1.65 per month for an effective pay-back time of about five years. Each recipient owns a tiny energy business.
- The U.S. PV industry would be given a solid jump-start; pull instead of push.
- Best of all, the American economy and conversation will be turned toward a sustainable, non-polluting, optimistic future instead of dependence on a waning oil age.
- A supplemental or companion program might be for our government to offer the same **U.S.-made panel to any U.S. citizen for half price of about $50.00 each**. This offer could be made on a one-per-person basis up to a specified annual maximum. At a million panels per year, the cost to the national budget would be fifty million dollars, a pittance considering the jump-start the program would make toward a solar-electric future.

How do ideas like these get traction? Only by grass-roots activism and exponential, self-feeding networking: **A "movement." It's up to you. Remember the two basic themes of this book: U.S. gasoline rationing and local lead-acid battery recycling are absolutely fundamental to an acceptable transition beyond the oil age.**

CHAPTER 6

Population and Per Capita Oil Consumption

BACKGROUND

This book joins a growing body of evidence that teaches we are at the peak or broad plateau of maximum world oil extraction. **This precarious ephemeral period in the epoch of recorded human history, itself only a tiny fraction of longer archaeolocal time, signals the end of energy-dependent growth.** Oil and the myriad petroleum-based products we have become so dependent on, in just the last century, are critical and fundamental to our modern lifestyle and all other energy sources. These undeniable facts are further exacerbated by a debt-based financial system, also dependent on continued growth and which cannot function without cheap and abundant energy, specifically oil.

This opening statement is ominous enough, but it does not include the concurrent three-fold explosion of world population also in the last century, a little over one human lifetime. It is totally incomplete to focus only on the contemporary peak and imminent decline of geologically-finite oil while human numbers continue to steadily increase. **The purpose of this Chapter is to develop a quantitative view of the second half of the oil age juxtaposed against various scenarios of continued population growth.** The specifics of peak oil are frequently marginalized or ignored by population activists and visa-versa. Both subjects can be lost in the drum beat of environmentalism, and climate change. Obviously, all are related and extremely complex. The main-stream public hears only confusion and non-quantitative panaceas.

POPULATION GROWTH VS. OIL EXTRACTION

Starting back with Figure 7 in Chapter 4, shown are eight different rates of population growth and possible decline by age and fertility rates ranging from three children by each female (3 cpf) to no children (0 cpf). Contrasted, in heavier lines, are three curves of projected world oil extraction. The basic middle curve "H" is the decline expected for the second half of the oil age and typical of any finite-resource extraction. This phenomenon for oil extraction is well established as Hubbert's Curve. The area under the "H" curve, beginning in 2012, now at or near peak, has been optimistically justified because of higher oil prices and more expensive extraction technologies.We are led to believe there are 1.3 trillion barrels of remaining oil. This prediction can be substantiated by the fact that **we have already used over one trillion barrels in the first half of the oil age** and the world extraction rate of conventional oil, as shown in Figure 3, has hardly increased since plateauing at 75 million barrels per day (or a billion barrels every thirteen days) in 2005. **In just the last twenty years, about one generation, the world extracted and consumed about half the trillion barrels used so far in the total oil age.**

The suggestion of an imminent end to the oil age is so ominous and alarming there is a strong counter-movement underway to debunk these numbers. Obfuscation has increased by the inclusion of non-conventional oil and other liquid fuels. We will never use the last barrel buried somewhere in the earth, but there is no denying that remaining oil is becoming increasingly expensive to extract, both energy-wise and financially.

Oil extraction

Curves (NO) in Figures 4 and 7 show the net oil available for use, **after steadily-increasing oil (or energy equivalent) input of one percent per year yields even less usable oil output than curve "H."** This is called Energy Returned on Energy Invested (EROEI) and shows a more accurate, but dire picture for the timing and availability of remaining oil in just the next 50 years. The third oil curve in Figure 7 (NO +1/2 trillion barrels) and shown by shaded squares, **gives a one-half trillion additional surplus, benefit of the doubt, to the optimists (energy "experts", politicians, and economists) who argue that improved technology and new discoveries will prove Hubbert grossly wrong.** None of the three curves shows any increase in the annual rate of world oil extraction, thus signaling an end to unprecedented oil-based growth as in the last century. The curves shown are for conventional crude oil and not liquid byproducts of natural gas extraction like "condensates" and natural gas liquids (NGL's). Non-conventional liquids like biofuels and tar-sands oil are not included because of their low EROEI and minimal effect on the conclusions.

Population growth
(new, unique, quantitative methodology, not extrapolations)

After spending much time trying to find, and/or not believing population projections that didn't make sense, I developed my own spread sheets to provide numbers I could trust.

Eight combinations of possible future population projections are shown in Figure 7. They were calculated using the following starting point and ground rules:

1. A "snapshot" of time (shown as 2012) defines the beginning point for each series of calculations.
2. Two demographic profiles are used. They differ by percentage of the total population divided into different age groups sometimes called cohorts. Obviously, if we start with a profile with mostly only young people just entering their reproductive age, there will be a much greater population bulge (total number moving forward) than if the starting profile is made up of only middle aged grand parents and seniors just beginning to die off. For my analysis I picked two different starting profiles; one from world, and one from U.S. census data. The younger demographic profile (ydp) is typical of the world as a whole. The older profile or distribution by age (odp) is for the U.S. To Summarize, the percentage of the total population by average age in each age group (cohort) is as follows and shown in the lower left corner of Figure 7:

	Younger Demographic Profile	Older Demographic Profile
Years 0 to 20	40%	32%
Years 21 to 50	43%	47%
Years 51 to 80	17%	21%

3. The average age at reproduction is 25 years old, that is if each female has only one child (1 cpf) that would be when she is 25. If she has 2 children, one might be at 24, and one would be at 26. Of course, in real life, there will be some females having children in their thirties, but this would be averaged by teen childbirth and the math and the conclusions would not change. Child gender distribution is assumed as 50/50.
4. The average age at death is 80 years old. Some may die at 65 and others at 95. The total, average numbers remaining to be fed until 80 would be the same. The luxury of modern health care leading to unprecedented old age

has been labeled "death control." At this point the crux of this entire population discussion will be emphasized one more time: **We can't simultaneously have both traditional fertility rates and modern old age.** A quantitative law jumps out: **It is numerically impossible for any closed society (no immigration or emigration) with a typical age distribution profile, like the world or even the U.S., to reproduce at a rate greater than an average of one child per female (1 cpf) and avoid increasing population in the near term future, if the members expect to live to be grandparents and great grandparents.**

It is true, and conventional wisdom, that a fertility rate of two children per female (2 cpf) will eventually level off at a "replacement" level. **But, as shown in Figure 7, this would take about fifty years and the final, stable, closed-society population would have increased by thirty percent and not decline thereafter.** We can't live to be old with modern healthcare and adequate food and concurrently have more than one child per female (1 cpf). Each of the increasing populace will be competing for a maxed-out food supply. It's been shown, historically, that it is impossible to support increasing numbers without, at the same time, inevitably degrading the agricultural base (carrying capacity).

History teaches of numerous "crashes," "collapses," and "overshoots." **This is already happening today in large parts of the world while, at the same time, we are leaving the artificial, oil-based energy level that made the excess population possible in the first place.** Our short oil age has facilitated old age in many ways; sharply reduced manual labor, dependable year-round nutrition, improved health care, and reduced infant mortality. We would all like this lifestyle to continue.

5. It should be clearly understood that the above ground rules, conclusions, and methodology hold true **regardless of the original, numerical size of the closed society.** I conveniently selected 7 on the digit scale, beginning at zero, on the left-side "y" axis to represent the world, the largest undeniably-closed society which has presently swelled to over seven billion people. All sub-societies as a part of the finite world must together average to equal the world growth or decline numbers. Some local societies may grow more, some less, but each will have to follow the same methodology. Some, like China, may attempt to take control of their population destiny. Others, like sub-Saharan Africa, will just let nature take its course and, without additional energy and food input from somewhere else, must suffer the inevitable consequences of exceeding their regional carrying capacity.

6. Another reason for using single digits on the "Y" axis is to show world oil extraction in tens of millions of barrels per day on the same graph. This shows clearly how population continues to increase while the temporary, two lifetime (eight-generation total) oil age is about to enter into its second half. All societal subgroups will experience the growing tension (gap) between increasing population and decreasing oil. An idealistic, "localized" community of seven-hundred, or an autonomous nation with seventy-million, even if each has bountiful food resources for their present population, must ultimately respect the same numerical limitations of reproduction.

When times are good, like any species, human population increases to the limits of carrying capacity. The excess numbers begin encroaching on their neighbors or suffer Malthusian "misery." All it takes is a climate event or poor land-use to trigger disaster.

The math

This section is included to show the methodology. **To repeat, the conventional wisdom that a "replacement" fertility between 2 and 2.2 will suffice is dangerously false. (Fast forward to the next section to avoid the details.)**

It is simple, but tedious, to do the numbers using the above ground rules. Examples follow so anyone can verify the population curves shown graphically in Figure 7; or use a different demographic starting profile than the two shown.

Referring to the younger demographic profile (ydp), typical of the world, we see 40% of the total population is in the cohort between 0 and 20 years or 2% per each year. Likewise, there are 43% /30 (divided by 30) or 1.43% each year between the ages twenty-one to fifty, and 17%/30 or 0.57% per year between the ages of fifty-one to eighty. At the end of the first year, each female who reaches 25 gives birth to her only child (1 cpf). Because 1.43% of the population is now 25 years old, the female one-half of this age group (1.43%/2) or **0.72% of the total population will be added as new babies.** For instance, if the closed-society population was 1000, 7.2 new babies would be added.

In this same first year, **0.57% of the population, male and female, would die** leaving a gain of 0.72% minus 0.57 % or a net increase of 0.15%. This does not sound like much, but the significant point is that the **population continues to increase even with a fertility rate of only one child per female (1 cpf).** As this birth minus death rate continues for four more years, the net gain in five years would be 5 times

0.15% or 0.75%. Our hypothetical population of 1000 has now grown in five years to 1007.5 people. If we did similar math for a fertility rate of two children per female, our community of 1000 would grow to 1043 mouths to feed in five years, hardly sustainable with a fixed agricultural base. Similarly, in five years, a world population of 7 billion people will grow by 301 million to 7.3 billion people.

When we enter the sixth year of our model, things get more complex because the original twenty year olds are now reaching the average reproductive age of 25, and there are more of them. At the same time the numbers reaching the average age of 80 are the same as the first five years. Using the younger demographic profile (ydp) typical of the world, there are 40%/20 years or two-percent per year of the total population turning 25 years old. Of these, one-half or one-percent are females. If each female has one child and die-off continues at 0.57% per year, the net gain each year is one minus 0.57 or 0.43% per year. This average pace will continue for the next twenty years until the last original female baby, less than a year old when we started twenty-five years ago, reproduces. At this rate in the next twenty years, population would increase another 20 times 0.43% or 8.6%. Our original hypothetical population of 1000 would add another 86 mouths to feed in addition to the 7.5 increase in the first five years, **for a total increase of 93.5 in twenty-five years, with a fertility rate of only one child per female!** With this model, the starting world population of seven billion will increase to 7.654 billion.

In the 26th year, for the first time, the female babies born in the first year become mothers, and our original mothers become grandmothers. Of the 0.72% of the population born in the year one, only one-half (0.72%/2) or 0.36% are females. If each mother continues the one child per female model, there will only be 0.36% of the population added as new babies while, at the same time, the death rate at 80 is still 0.57 % per year. This contrast finally leads to a negative population growth of 0.36% minus 0.57 % or a negative 0.19%. per year. For the next five years this downward pace will result in a total population reduction of 0.95%. Our original community of 1000, that had grown to a peak of 1093 will now begin to decline by the year thirty to 1084 people. For example, a **world population starting at 7 billion would pass maximum population and then decline to 7.588 billion, but it took thirty years even at only one child per female!**

Conclusion

By now, the mathematical methodology presented should be clear. In the graph (Figure 7) the numbers are continued for 80 years, two different starting age profiles, and five different fertility rates. The conclusions are profound and disturbing.

Any closed community, nation, or world, living near the limits of its diminishing finite resources, specifically, energy and fuels, non-renewable minerals, arable land, and water, cannot reproduce more than one child per female, and simultaneously live to old age without exceeding its carrying capacity. To repeat: a society can't have it both ways! This conclusion may, in a nutshell, be a short history of, and future prediction for, the world including all closed-loop, smaller communities. Constant ignorance and violation of these basic mathematics have repeatedly led to starvation, war, genocide, ecological devastation, infanticide, deprivation, misery, and even cannibalism. The entire world is now entering this tipping point. Unrest in many parts of the world including Africa, the Mid-East and American inner cities are clear manifestations of undeniable mathematics.

PER CAPITA WORLD OIL CONSUMPTION

A simpler, conclusive way to combine the contrast and imminent divergence between increasing population and concurrent declining energy is to divide the total world oil consumption by total world population. **Even one barrel of oil per capita would be completely game-changing in the long history of human and animal muscle power.** A single barrel of oil contains 42 gallons of extremely convenient stored energy. This incredible amount is equivalent to 1,384,000 watt hours or 4.7 million BTU. It would take a human working **continuously** eighteen thousand (!!) hours to generate an equivalent amount of energy. Admittedly, there are efficiency losses in converting the oil-energy to mechanical work or heat, but an oil-powered machine does not need to eat or stop for rest.

A pint or two of equivalent liquid fossil fuel energy would have been utopian to pre-industrial humans. Now we take for granted thousands of gallons a year, for each of us; convenient labor-saving energy but at wildly different rates throughout the world. **The per capita curves in Figure 8 demonstratively show the rise and fall of the oil age in a two-lifetime span of 160 years. There was a little oil before this time span and there will be little left afterwards, so for all practical purposes, the total oil-age will be very short.** Chapter 1 and Figure 1 introduce this per capita analysis, especially the highly-skewed contribution of U.S. gasoline.

The demographic methodology for population momentum shown in Chapter 4, Figure 7 shows that the continuation of a **stable,** per capita oil-energy availability with **any fertility rate greater than zero (0 cpf) cannot coexist with the expected decline in oil extraction ... no matter how optimistic the decline might be!** In the following Figure 8, the present world average, per capita extraction and

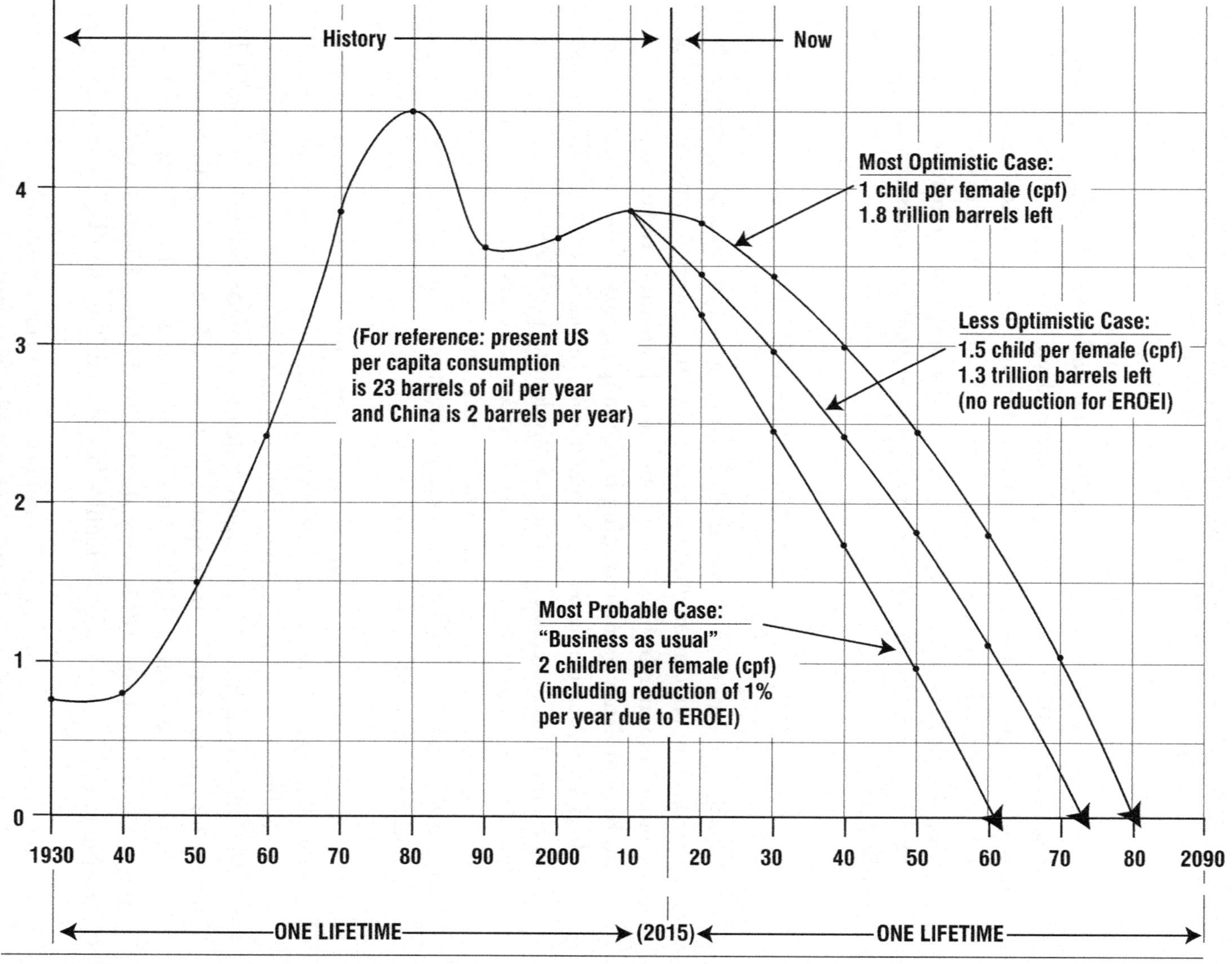

FIGURE 8 Per capita world oil consumption in a two lifetime span

consumption of less than four barrels per year, per human, is beyond peak by over thirty years. Currently the most energy extravagant consumers, led by Americans, each presently use (burn up) over twenty-two barrels of oil per year, six times the world rate. In fact, Americans use about twenty-five percent of the world oil production with only four percent of the population. China is gaining but still uses only twelve percent of world oil with twenty percent of the population for a per capita consumption of two barrels per person per year, about one half the world average shown in Figure 1.

U.S. oil extraction and consumption

Because there is only one world, overall extraction and consumption have to be the same. Therefore the per capita average is simply the ratio between usage and world population. However, as shown in Figure 1, there are gross disparities between different nations that constitute this average. Some parts of the world have achieved exceptional consumption rates typified by the lifestyle we Americans take for granted. The references for the numbers are U.S. census data and the EIA (Energy Information Administration).

The next graph, Figure 9, shows the historic first half and the projected second half of the two-lifetime oil age for the U.S. lower 48 states and Alaska. Also included are U.S. population numbers also reported by the census department prior to 2010 and projected forward using the methodology explained earlier in this chapter. Also, as a subset of the world, we in the U.S. have to deal with immigration and emigration adjustments. The fertility-rate population projections and conclusions would be the same for any autonomous nation. **It makes no difference if the population is immigrant, white, black, religious, rich, poor, republican, or democrat. Everyone has to eat. My analyses are intended to be absolutely apolitical.** There are no reasons why progressive or conservative mathematics are any different. Numbers have no party affiliation.

A most important observation in Figure 9 is that, U.S. extraction did, in fact, peak in 1970 at 3.5 billion barrels per year or 10 million barrels per day. This is about one half the present consumption rate of 19 million barrels per day, down from 20 several years ago. M. King Hubbert was publically derided for predicting the U.S. peak back in the 1950s. But subsequent reality proved the veracity of his techniques which are now the context for prediction of the peak of world oil, and the second half of the oil age with the following provisos:

U.S. oil **extraction has not** exactly followed a mirror-image decline rate as would be simply predicted by a Hubbert's curve. New technology and eight-fold higher

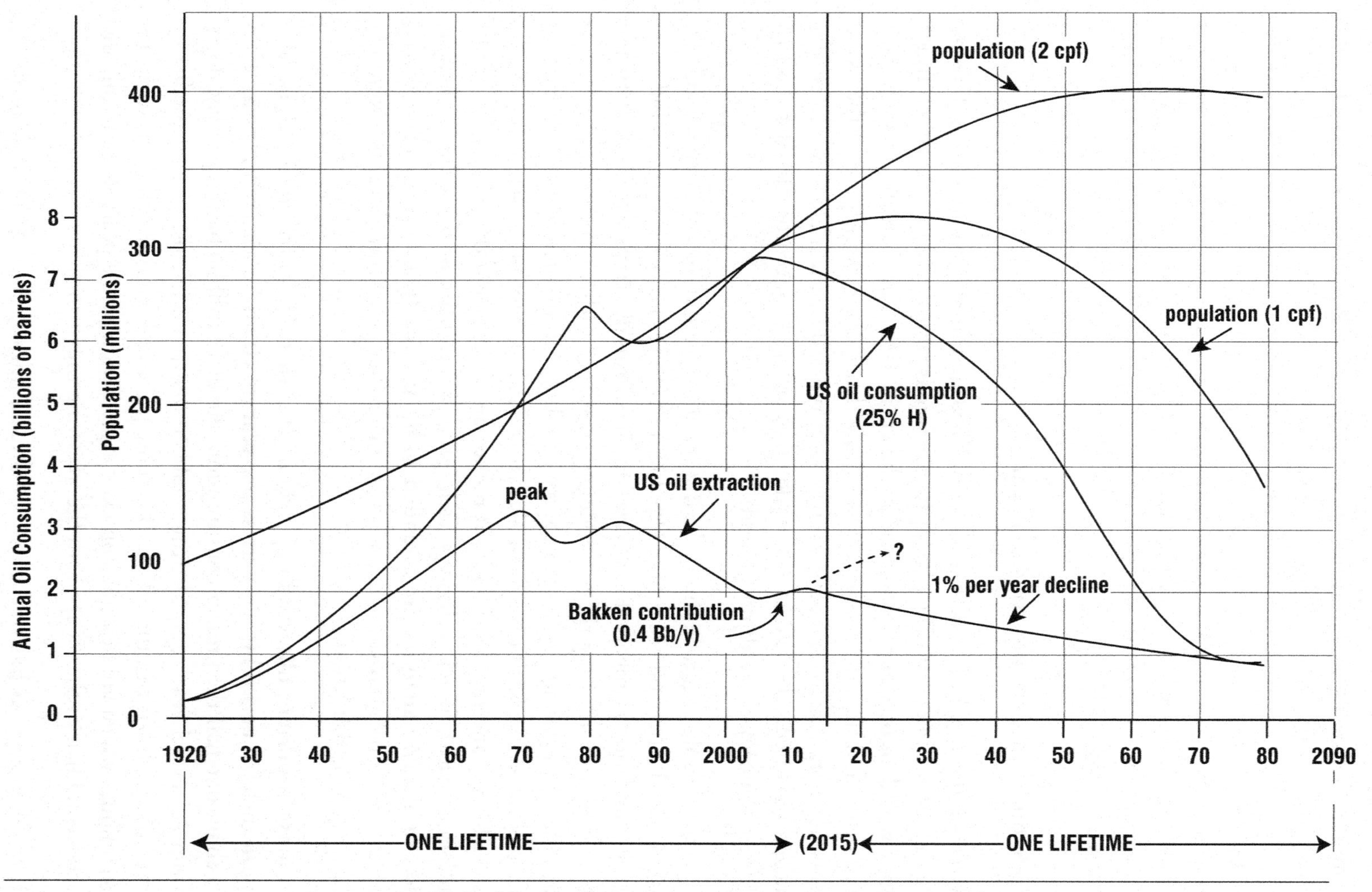

FIGURE 9 United States oil extraction, consumption, and population

prices have eased the decline rate somewhat, but there is no argument that domestic extraction of conventional oil declined by 2005 to half of the 3.3 Bb/y (9 Mb/d) it was at peak. During the rapidly growing period from 1920 to 1970, oil extraction increased by about four-percent per year. After peak in 1970, U.S. oil extraction decreased about one and one-half percent per year. By 2014, U.S. lower 48 states plus Alaska extraction climbed back to 2.5 billion barrels per year or thirty-six percent of our consumption of 7 billion barrels per year.

This is hardly "game changing" or a "revolution" as touted by the "Wizard of Infinite Oil," Daniel Yergen in his February 5, 2013 testimony to the House Energy and Commerce Committee. The legislators and American public do not comprehend the magnitude of these numbers and are led to believe that a resurgence of several million barrels per day will prevent the end of the oil age. In order to give every benefit of the doubt to these optimists, the reduction of future US **extraction** rate is projected (flattened) to one percent per year.

U.S. oil **consumption** (including, but far more than U.S. oil **extraction**) is continued forward in Figure 9 as one-fourth of predicted world consumption. Steadily declining EROEI is not included although oil will continue to become more and more difficult to extract. **Also, in favor of the optimists for non-conventional oil, an additional 300 billion barrels (0.3 trillion) are added beyond that expected in the second half of Hubbert's curve of world extraction.** Continuing forward just 58 years to 2070, U.S. oil consumption would, as one-fourth the world extraction rate, drop below the more-optimistic U.S. extraction decline rate of one percent per year. By that time, total U.S. consumption would have to **decrease by over eighty-five percent** and by then, in less than one lifetime, we either would have taken extreme measures to decrease **both** population and energy consumption or total chaos will have ensued, nationally and world-wide. (See the first four scenarios in Figure 2.)

Figure 9 also shows future U.S. population momentum at one child per female and two children per female using the same methodology (including the eighty-year life span) explained earlier in this chapter. (Present U.S. fertility is 1.86 cpf). **Note again (and again!) that the conventional wisdom of two children per female "replacement rate" further exacerbates our predicament by increasing population thirty percent higher than now, before stabilizing and not dropping.** In the last few years, in rounded numbers, the U.S. annual net increase is four million births **plus** one million immigrants (one-half illegal), **minus** two million deaths, for a net increase of three million per year (4+1-2=3). **That's 250,000 per month additional jobs and food required just to keep from slipping backwards to more unemployment and economic decline.**

Per capita U.S. consumption

The next graph (Figure 10) combines the extraction, consumption, and population curves developed in Figure 9. At the bottom for reference is the "less optimistic" average world per capita oil consumption for 1.5 children per female. The much higher U.S. consumption of 25% of world oil is projected forward on the same time-line at one and two children per female. Now we can see that 1 child per female gives about a ten year delay before reaching the lower per capita consumption as would be expected with 2 children per female. **The continued attempt to access 25% of world oil ensures we will continue our part in competition for dwindling oil including directly-related foreign presence, geo-politics, territorial conflicts, and deteriorating human-rights.**

We can also see in Figure 10 that if the **U.S. had to depend only on domestic oil (the lower 48 states plus Alaska) for energy, we would already be closer to the four barrels per year current average for the world.** This supports the arguments presented in Part I.

Americans are precariously dependent on foreign oil, including Mexico and Canada, to continue any semblance of our unique energy-intensive lifestyle for a few more years. **A few extra billion barrels from ANWAR, off-shore, or tight "fracked" oil will not change this dire prognosis**. We will soon be forced, whether we like it or not, to sharply reduce our use of oil, especially for non-essential and inefficient transportation. In Chapter 7, the argument is made for coupon gas rationing as the only way to quickly begin an equitable reduction in oil consumption regardless if it comes from domestic or foreign sources.

As the oil age winds down, our food system will be seriously impacted because **it will no longer be possible for one farmer to feed 300 people thousands of miles away.** To repeat the continuing theme of this book, the best we can plan for is that the **U.S. carefully nurtures its remaining oil endowment by drastically reducing consumption, reduce fertility rates, seriously limit immigration, move to local or personal agriculture, and begin immediately to segue to a more expensive (than fossil energy) solar-electric future.** Of course, none of these are likely to happen, let alone all in tandem, without a grass-roots education effort, "It's Up to You." Although the whole subject of transitioning to a post-oil age seems hopeless does not mean it should not be clarified, quantified, and publicized for reference in the near-term future as reality becomes apparent.

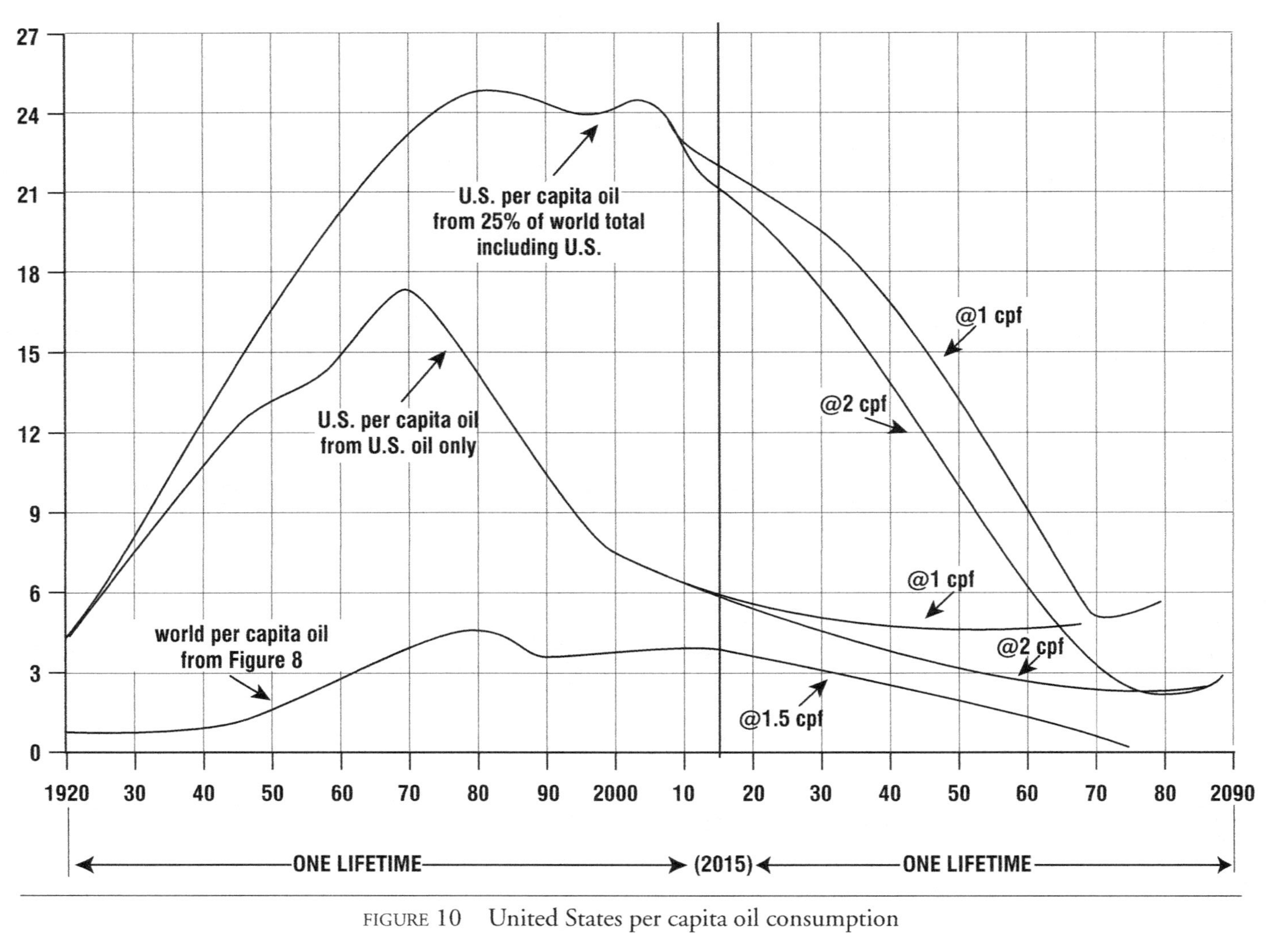

FIGURE 10 United States per capita oil consumption

The only answer

To repeat, it should be clear that the only way Americans could significantly mitigate the end of the oil age beyond one lifetime is to sharply reduce per capita consumption from our present profligate level of twenty-two barrels per person per year to the world average of four as explained in Chapters 1 and 2. This is why the low-energy, solar-electric system described above must quickly extend and overlap waning oil. This is why we must begin to ration oil consumption, starting with gasoline, so everyone has to pitch in. If we were to reduce U.S. per person consumption to the world average of three barrels per person per year, and with our population growth reduced to a fertility rate of 1 child per female, scenarios 5 and 6 in Figure 2 show a U.S. oil age extending between 55 years (still less than the lifetime of a child born today), and possibly much longer. The difference, of course, depends on how much (and at what price) domestic oil remains and how carefully we nurture what's left for our descendants.

THE HISTORY of BIRTH CONTROL and OTHER VOICES

The ultimate societal challenge through the ages has been to control and reduce human reproductive rates to a level commensurate with finite and now-declining energy supplies. This will take massive education, peer pressure, and honest leadership within a closed sovereign national or regional society. Everyone must be involved. Those who are not, only exacerbate the predicament, guaranteeing that all will be dragged down together. For a regional society to succeed in rising above a larger collapsing world civilization, it will also have to be diligent in controlling immigration from envious foreigners.

The bibliography at the end of this book includes many titles specific to population; starting of course, with Thomas Malthus who has been almost forgotten in the last two-hundred years because of new lands, high-tech agriculture, and most significantly, the sudden utilization of fossil fuels.

Modern contemporary authors are Paul Ehrlich who wrote the best seller *The Population Bomb,* and Meadows et. al. *The Limits of Growth*, first published in the seventies. Since then, both these authors were "proven wrong" because the vast potential of the world's finite resources and the technology of the "green revolution" had yet to be fully realized. Now, their voices are ringing true. More recently is the extensive body of work by my good friend and mentor, the late Albert Bartlett summarized in the book *The Essential Exponential.* He has given thousands of lectures throughout the world about the mathematics of population growth. These can be found on youtube.com.

Another contemporary author who integrates both sides of the population-resource equation is Lindsey Grant; *The Collapsing Bubble* and *Too Many People.* He is a contributor to the U.S. non-governmental organization (NGO) Negative Population Growth (npg.org) which focuses directly on the subject. Another U.S. NGO is World Population Balance (worldpopulationbalance.org). I have an older book published in 1980 by M.Bayles *Morality and Population Policy* which raises the obvious question of why we should not have a moral obligation to future generations to leave them a tolerable world-balance of resources and consumers.

From the UK, several of the best references are: *The Rapid Growth of Human Populations* by William Stanton, and *The Growth Illusion* by Richard Douthwaite. Similar work is spearheaded by the Optimum Population Trust (populationmatters.org)

The recently published book *Countdown* by investigative reporter Alan Weisman is an amazingly comprehensive, first-hand, 500 page summary of world population problems.

The only possible way to achieve 1 cpf in a modern free society is with vast education, publicity, and peer pressure. **The public must realize that every child born today will not only compete with everyone else for resources, but their parents will still be here to suffer with them in a world becoming much more difficult.** Isolated bunker mentality will not survive the coming tsunami because of the limitations of localization as discussed in Chapter 9.

Blame the messenger?

I often wonder why I continue this unpleasant mission; but always come back to the conclusion that **our best gift to those already alive is to define how we could extend and possibly supercede the oil age by quantifying the facts and solutions, including fuel rationing and population control.** The weatherman's job is to foretell the future regardless of how unpleasant, so listeners can plan accordingly. We don't blame or ignore the forecaster.

Taboo subjects: sex vs. population?

We are a society saturated with sex. We joke about sex and are constantly bombarded by advertising, clothes, and entertainment in the same mode. **But, we dare not discuss "private", personal decisions that question family planning and more mouths to feed.** The innate desire to mate is one of the most dominant drives of any species, especially among males. Attracting mates, fleeing from danger, accessing food, and protecting or raiding territories are all genetically "hard-wired" in males for the perpetuation of the species.

Most males have no concern for the numerical outcome or future for the union between thousands of spermatozoa and millions (in the case of fish) female eggs. A few centuries of "modern man" cannot erase eons of successful competitive survival. It takes hundreds of generations for the genetic code to slowly adapt to changing environment, but the basic impetus to mate is always lurking, only recently tempered by the social mores of civilization. This may be why individual or team contact sports from gladiators to football, or violence thrive as entertainment. **They satisfy the basic urge to conquer the other guy and bring home the spoils to a cheering family.** Alternative sports which test individual speed of travel, or compare competence in overcoming natural adversity seem better than beating-up on each other. Modern gun culture may bridge both genetic drives: who is the best individual shooter … inferring the best chance in accessing food, or who can defeat (kill) the other competitor?

We might summarize this line of reasoning with a hypothesis: "higher intelligence" is a prerequisite for population control necessary for a long quality life in a closed society staying within the limits and carrying capacity of its own indigenous fragile resources. The antithesis is "less intelligence"which breeds greater numbers and thus overwhelms the visionaries (and resources) who are striving for long-term sustainability … higher intelligence loses, less intelligence wins, overshoots, and collapses!

In our great wisdom, we neuter our pets and cull our farm animals to restrict population within obvious limits. But now civilization is at the threshold of catastrophe because the decline of fossil-fueled food, and the longer-term context of climate change. There are those who argue that if we all live and eat like marginal third world people, or invent new ways to manipulate agriculture, we will be able to feed the nine billion mouths projected by the UN in 2050. Assuming these alternatives are possible and the crowding is acceptable, what comes after 2050, one half a lifetime from today? Sooner or later, any growing population must respect the limits of earth's finite carrying capacity. If not the result, as Malthus named it, is "misery." We can argue that peak oil, peak food, and *Peak Everything* per Richard Heinberg's book are here, now, but are we ready for "peak sex"?

One child per male

I for one am all for continuing sex, but not population growth. In my opinion, the better way than one child per female (1 cpf) is to switch the gender responsibility for the future of humankind to one child per male (1 cpm). In modern times, the answer is so simple: after his only child, every male gets a (free?) vasectomy. After an hour in the Urologist's office and a couple of days of discomfort, life goes on … without social disruption, abortion, abstinence, frustration, unpredictable birth control, unwanted children, and extra mouths to feed. **In addition, the single child will be the focus of all the love, attention and resources from both of his/her parents in the near-term challenging future.** I bet most forward-thinking females would welcome the idea.

Sounds simple but, like gasoline rationing, the devil is in the details. We still have to respect the traditions of religions as with Amish and Catholics. Will they out populate those making a conscious effort to provide a better life here and now, for us and our descendants? Who is sacrificing what, for whom? Low birth rate is already a fact; nearly down to 1 cpf in some countries like Italy, Japan, and Russia. Can 1 cpm be enforced as by law, peer pressure, or tax credits? The vasectomy can be reversed if the single child dies. At least the topics of 1 cpf and 1 cpm should be openly debated as in China and Japan for thousands of years. The autonomous groups that best resolve the declining-energy, increasing-population paradox will prevail in the coming years.

Scale of implementation

Following are five levels of human living-arrangement from the scale of the entire world of seven billion, down to the individual and/or including the immediate family of, at most, a few dozen people. I've listed my opinions regarding the chance of success for each to survive the population-resource challenges we face:

1. **World (global): There is no hope for the world with more than seven billion humans to reach and enforce any mutual agreements regarding resources and population.** There are far too many differences in religion, culture, proximity to each other, natural resources, language, and wealth... all interacting with the innate tendencies for procreation, survival, and greed. Suggested reference: Dawkins, *The Selfish Gene.* Attempts to prove otherwise like the U.N. are tentative and ineffective. Wars are the results of disagreements and differences. **A world-wide plan for rationing** or *Depletion*

Protocol (see the book by that name by Richard Heinberg) offers little hope for bridging disparate international interests.

2. NATIONAL: **Individual, autonomous nations like the U.S., or several with close cultural, resource, and language ties, are best suited to survive a post-peak-oil future.** The U.S. combined with Mexico and Canada would fit this category. They are large enough to have substantial natural resources, shared national security, food supply, crop diversity, complex manufacturing, and resilience to climate change. Each could ration critical resources with or without broad international cooperation. Singular nations, or close coalitions will have to defend their sovereign borders within common topography to control immigration. They have a better chance to integrate the knowledge, resolve, and support of the majority of their populace behind effective or common leadership. Physically, this intermediate size will be better prepared to survive a low-energy future especially with sharply curtailed travel and long-distance trade.

3. REGIONAL OR STATE: **Smaller segments of a national level have less chance to survive alone.** Although they overlap in history, language, inter-family relationships and closer travel; they are too small to be energy, regional weather, high-tech product, and food independent. Borders cannot be defined or defended. To do so would limit the flow of goods and people, usually with common goals and traditions.

4. LOCAL: **The potential for unique, isolated, long-term survival, or resilience in a localized community of a few thousand people is nil** because the autonomous group is too small to independently maintain a complex modern lifestyle. Although the movement is laudable because of the shared sense of security and temporary buffer against collapse, a localized community cannot stand alone for long and should not shield its members from concern over the larger picture. Ancient tribal patterns were stronger when extended to the limits allowed by topography and travel. This important sub-subject is expanded in Chapter 9.

5. PERSONAL: **This level, including immediate family, infers individual control over one's fate which is totally unrealistic.** Primitive survival required the support of at least a village, with success intimately reflected by the sum and average of individual actions.

CHAPTER FINAL THOUGHTS

The critical subject of population control has been addressed since the beginning of recorded history, and probably before, but not documented. All ancient, autonomous, surviving cultures had to thread a precarious fertility rate path of about three (3 cpf) to ensure continuity, but not grow depending on local, short, hard-life expectancy. Any more would have exceeded the local carrying capacity and lead to collapse. This was before modern health care and longer life spans. Now, we urgently need to reduce fertility to, at most, 1 cpf in order to navigate the end of the oil age ... including the ecological devastation we have caused by over populating every niche of the world. There are hundreds of books that venture into every detail of fertility control. I will not go further with this except by suggesting that male vasectomy would be a better choice rather than leaving it to women to be responsible for limiting population within available resources.

This solution seems far more palatable than severe local traditions as described in Arthur Boughey's 1976 book: *Strategy for Survival, an Exploration of the Limits to Further Population and Industrial Growth.* Examples like female infanticide or having to kill another person before marriage helped isolated island societies cope with population control.

Another of the best books in my collection on population is *More* by Robert Engelman, Island Press (2008). It is easy to read and explains older birth control methods like emmenagogues and pessaries (neither word is in my pocket Webster's Dictionary), which were used by women before modern contraceptives were invented. Bill McKibben wrote the book, *Maybe One.* An excellent read is *The Fatal Inheritence* by John Bligh, Athena Press, (2004). For a world-wide web site that advocates zero children see: vhemt.org. Very interesting.

However, in my opinion of comparative time frames, running out of oil is far more urgent than the directly-related crises of longer-term population control and climate change.

To know and not act is to not know.

(Chinese proverb)

CHAPTER 7

Gasoline Rationing, the Only Equitable Way

WHY RATIONING?

As an immediate antidote to the complex, intertwined, global, economic-energy crisis, **the United States Congress should legislate the reduction of gasoline consumption by some form of national coupon rationing.** This is a far better alternative than burning the remaining oil as quickly as possible, relying on wild market speculation, high-price polarization between rich and poor, or increased taxation (still another way of catering to wealth disparity). Gasoline rationing is the best move we could make to stem the hemorrhage of, at least, our own domestic oil.

At first this initiative seems counter-productive for a sluggish economy, but further examination shows why we need a bold about-face in our thinking and lifestyle, instead of attempting to continue business-as-usual as we enter the second (declining) half of the short two-hundred year oil age. **Today, Americans alone burn-through close to four-hundred million gallons of gasoline each day, an economic drain of one billion dollars at the depressed price of $2.50 per gallon.** Every gallon of gas consumed represents precious finite oil gone forever, instead of being available in the future for every link in the long food chain, essential transport, infrastructure maintenance, accessing other energy sources, national security, or providing feedstock for thousands of other materials. Our and our children's survival are being sacrificed to today's profligate waste. Figure 1 (page 2) shows that U.S. gasoline consumption alone is 9 million barrels per day or almost 11 barrels per person per year. The display in Appendix A reminds us of the many ways that oil, including diesel and gasoline, is woven into our modern lifestyle. Some are far more critical than others. For example, food production and oil-support for the production of other lesser energy sources are far more essential than petro-fueled entertainment or frivolous travel.

Gasoline rationing would help mitigate and insulate the U.S. from the inevitable post-oil age collapse of industrialized civilization by conserving our domestic supply. As disconnected as they may seem, closer examination will show why the decline of finite fossil energy (beginning with peak oil), a debt-based financial economy, and gasoline consumption are closely interactive. If the case for gasoline rationing seems preposterous, unnecessary, or unworkable, at least skip to the summary review and conclusions. Retain and refer back to this chapter for reference in the next several years after the failure of all other attempts to continue high-energy travel such as electricity, hybrids, biofuels, **or the myth that U.S. extraction rate will double back in four years to its peak in 1970**. None can rescue us from our love affair with gasoline-powered personal travel.

A brief history

Gasoline rationing is not a new concept. Typically, commodity rationing is a **temporary** expedient necessary to ensure equitable distribution of fuel or other needs in times of scarcity. As a boy, I remember gasoline (and food) rationing during WWII. These were perceived as precarious times and everyone pitched in to be sure no one was left out, or inflation restricted access to just a few. More recently, there was a faltering attempt to ration gasoline during the Arab oil embargo in the early 1980s. But this crisis quickly passed and the world continued on a path of inexorable growth fueled by plentiful conventional oil gushing from the North Sea, the North Slope, Mexico, Russia, the Mid-East, and Africa. This pattern continued through the Reagan/Thatcher years and any thoughts of oil (or gasoline) shortages were quickly forgotten.

A "depletion protocol"

During this remission, into the 1990s, new voices lead by Dr. Colin J. Campbell, a British petroleum geologist, warned that the sum of all world oil-extraction rates would eventually peak and the industrial age would soon enter a new phase of contraction. **This time, the decline of oil would not be a temporary scarcity, but instead, usher in a permanent and drastic change in the history of the world.** Campbell's work evolved into the world-wide Association for the Study of Peak Oil (ASPO) with many sub-chapters in different countries; for instance, ASPO-USA (.org).

(When this final book is published in 2016, Dr. Campbell has kindly offered his remarks in the included forward.)

As the quantitative facts regarding oil extraction rates and country-by-country depletion became more accurate, it was obvious that some type of world-wide oil consumption-curtailment would be best to match the imminent decline of world-wide extraction and avoid plunging the world into chaos. A comprehensive proposal to reduce consumption rates in synch with declining extraction rates was first proposed by Campbell as the Rimini protocol and then changed to an oil depletion protocol (ODP). This proposal was, in effect, a world-wide rationing plan meant to include all user and extracting countries. An excellent summary of this work can be found in Richard Heinberg"s book, *The Oil Depletion Protocol: A Plan to Avert Oil Wars, Terrorism, and Economic Collapse* (New Society Publishers, 2006). The sub title refers directly to the collateral damage caused by oil scarcity and increased cost.

Unfortunately, now in ten-year's hindsight, it is obvious that an international (or national) rationing plan could never be implemented for at least five reasons:

1. The combination of high cost, resultant "demand destruction", and enhanced recovery of non-conventional oil has kept us on a plateau of consumption for the last ten years.
2. The ODP was intended to be world-wide. As we enter the post-oil age, and similar to the Kyoto environmental protocol, it will be every country for itself. There is little hope for international cooperation on any issues, especially energy, climate, and population control.
3. Even if downsized for the U.S. only, the ODP attempted to include all oil uses, not just gasoline.
4. The ODP was too complex and impossible to administer globally.
5. The public is continually reminded by well-heeled energy-funded advertising that there will always be plenty of oil and/or easy substitutes. "Trust us. Don't worry, the Peak Oil theory is dead."

There is a new book, *The Impending World Energy Mess* by several ASPO-USA members and energy experts , Robert Hirsch, Roger Bezdek, and Robert Wendling. (2012 apogeeprime.com). A considerable portion discusses the merits and "complications' of gasoline rationing as a possible "mitigation" for the post-oil age.

Another excellent history and case for fuel and food rationing can be found on Sharon Astyk's blog site: scienceblogs,com/casaubon's notebook/2010/08/24. Sharon was a fellow ASPO-USA member and to quote from her comprehensive ten-page paper: "Rationing is both possible and potentially quite palatable, as long as it occurs in the context of public education and strong connection to current events."

AN ENERGY BALANCE

In late 2008, the world slipped into an economic recession exacerbated by the tension between the need for continued growth, but constrained by finite or declining resources; specifically, conventional oil extraction. **The basic absolute prerequisite (along with non-energetic raw materials) for economic growth, travel, and transport is readily-available energy.** This premise is expanded in Chapter 10. Beginning in about 2004, the world-wide extraction and consumption of finite oil, began to plateau and faced an unprecedented terminal decline. Superimposed on this geophysical reality is a debt-based economic system, predicated on continued future growth, and therefore can no longer continue. The geological limit of liquid, pre-stored energy dictates the rules we must obey. Temporary reprieves from new ways to search farther and deeper, or more-efficiently stretch the energy we have, only postpone the inevitable shortfall by a few additional years.

The Energy Barrel

A simple analogy of energy flow into and/or out of a storage reservoir exemplifies our predicament. This concept is shown in Figure 11. The concentration and storage of energy in any form are always difficult. Ancient sunlight-energy, stored as any of the three fossil fuel hydrocarbons, is by far the best way ever known. Energy is that elusive multi-faceted capacity of something to do any combination of work, heat, and growth. As with fossil fuels, wood, or food, energy can be stored in a real tangible substance. Energy should not be confused with currency and wealth, which are only convenient substitutes for conveyance or ownership of physical energy in many different forms as long as the system is stable and a quantitative equivalent value is agreed-upon and respected. But, without continued availability of readily available real energy, not a currency (money) surrogate, growth cannot continue (see following discussion of "false" energy inputs). A dynamic system will contract and cool if more energy is lost than is replaced. **In its most basic form, any real material thing, including its energy content, must first satisfy the essential requisites of life; that is, feed you, move you around, or keep you warm, otherwise it very quickly becomes lesser in importance and value.**

For easy analyses, scientists and engineers often avoid the internal, dynamic energy-interactions in a complex system by analyzing and quantifying **only the external ins and outs of energy over a period of time**. This avoids having to micro-analyze minute details inside the system. **Since energy cannot be created or destroyed, the difference between all the energy inputs, minus all the energy outputs,**

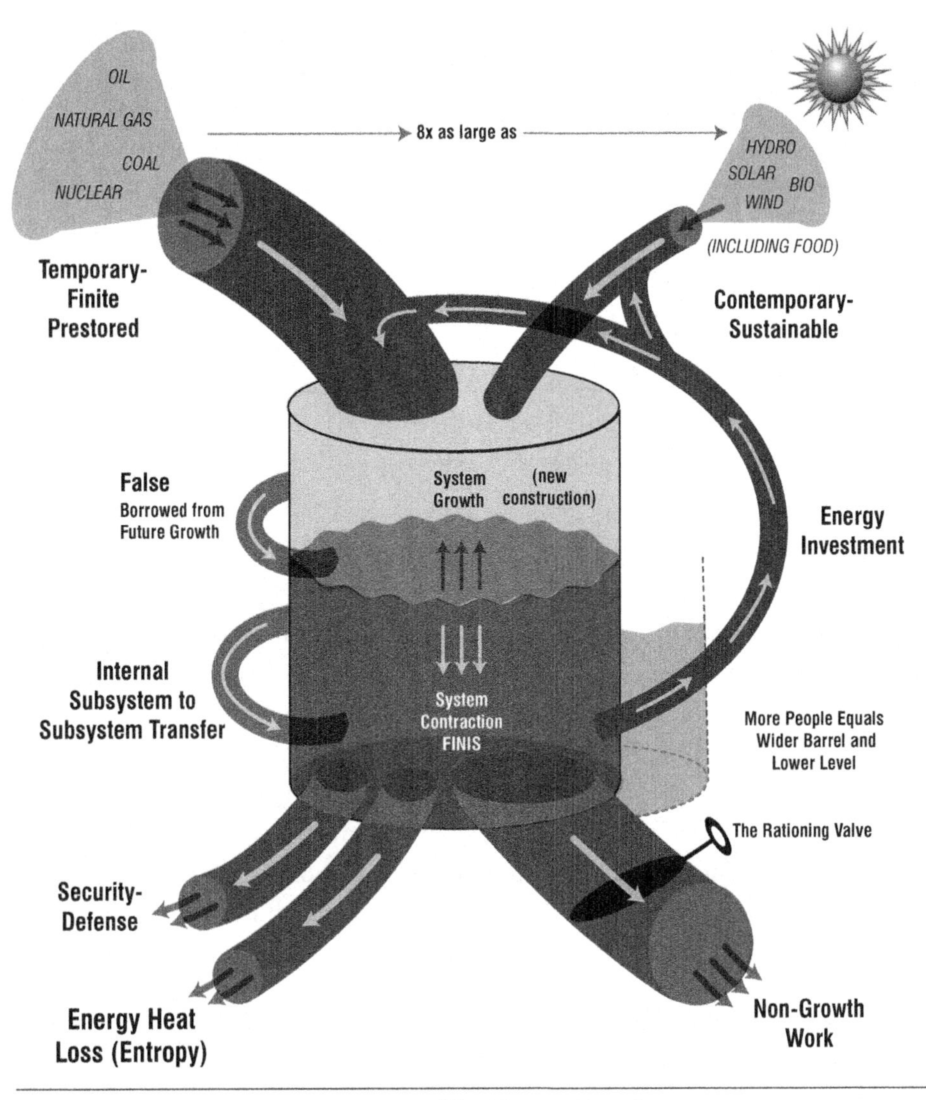

FIGURE 11 The Energy Barrel

must be accounted for as internal system growth, temperature change, storage, or contraction (decay) and can be summarized in an energy-flow model as in Figure 11. If temperature remains stable, in minus out equals growth or contraction.

This basic concept is no different than water flowing into and out of a barrel or electrical current flowing into and out of a storage battery which can, independently and sporadically, feed various loads. Another analogy is the amount of energy from food flowing into our bodies while energy in the form of work and/or heat, to maintain our metabolism and temperature, flows out. An excess of inflow over outflow, including waste, appears as growth (even obesity).

Yet, a misunderstanding or disrespect for the balance of energy in a finite system underlies many seemingly dissimilar problems like running out of gas, excess weight, economic recessions, or the entire collapse of civilizations. Understandably, much confusion arises because energy can hide in many different forms and is always trying to dissipate to a lower-energy form (a process called entropy). For instance, the barrel can leak, a warm body may cool, the water can evaporate, or we will starve for lack of food. Energy is the fundamental prerequisite for life and movement. Power is not synonymous. Power is only a measure of how fast the energy is added, dissipated, or is changing in form inside the system as explained in Appendix B.

Specifics (energy inflow)

By far the largest and most important factor either in or out of the model in Figure 11 is the overwhelming contribution (about 85%) from TEMPORARY-FINITE fossil fuels. Nuclear is also a smaller finite energy factor (about 6%) because fissionable uranium, which peaked in production about 1980 during the cold war, is also non-renewable. TEMPORARY-FINITE inputs must eventually and inexorably diminish to zero, by their very definition as finite. Internal system growth will cease and begin to decline **unless energy outputs also proportionally decrease, or other inputs increase enough to make up the shortfall** of TEMPORARY-FINITE energy and keep the entire system from contracting and soon collapsing.

We waste much valuable time trying to find new energy sources, arguing about how much is left, or how much the climate is changing while, in 2016, non-conventional oil extraction will begin to seriously decline because of low oil prices. All finite fuels (natural gas, coal, and fissionable uranium) will soon follow by 2035 at the latest (see Figure 6). We have clearly entered the second half of the brief industrial age. As would be expected, the first half provided the basis for un-precedented growth, including human population. **The second half will lead to oil (and total energy) contraction tempered by how well we restrain the outflow losses of the energy-balance system.** Since food is highly dependent on fossil energy there will be critical constraints on the human population that can be fed. Newborn babies will grow to be consumers in the same time frame that input energy decreases. The cover of this book shows this pictorially.

A significant concept, which should not be misconstrued as an energy input and which leads to economic crises, is the FALSE borrowing of wealth from the future based on the expectation and perception of continued growth of the complete system. **As explained earlier, wealth, as measured by paper currency or metal coins, is not energy. It can't do work or keep us warm.** It is only a temporary quantitative value assigned to real things like energy, food (another form of energy), goods, or services. Non-energetic wealth only represents the ephemeral, peaceful agreement of its contemporary worth as a surrogate for "real things." It is subject to inflation or collapse at any time.

Continued growth infers an increase of wealth as an indicator of additional "real things" and is equal to the original quantity (principle) plus interest, hence, "real growth." But real system growth **cannot** happen without a steady surplus (input over output) of either dependable TEMPORARY-FINITE energy inputs pre-stored from the past, or an increase in CONTEMPORARY-SUSTAINABLE inputs from the present.

Unfortunately, the borrowing of future wealth based on extrapolated past trends and expected continued growth is the backbone of our entire financial system. **But remember the basic premise: A system cannot grow without ever-increasing energy.** It may take time for this connection between energy and wealth to make sense. But think about it, you can't refill a pail of water today with next week's rain. You can't satiate today's hunger by thinking about next week's groceries or planning next year's harvest. Chapter 10 continues the discussion of the economics of energy.

The leveling or peaking, beginning in 2004, at about seventy-five million barrels per day of conventional crude oil production, and resultant steady price increase led to economic stress and decline, first of the growth-related housing and then of the entire, intertwined, debt-based, highly-leveraged, world-financial system. Everyone in the world started to spend more of their wealth on energy including all energy-related goods and services. This left less for non-energy spending and overall system growth.

By the last quarter of 2010, **the total of all liquid fuels** including non-conventional oil from tar sands, tight oil (fracked), deep-off-shore oil, polar oil, natural gas liquids, and bio fuels reached eighty-five million barrels per day. The total increased another six percent to over ninety million barrels per day by 2014 (see Figure 3, on page 35). The diminishing energy return on energy invested (EROEI) and much higher cost for non-conventional liquid fuels exacerbate the tension between higher extraction (production) costs and continued population growth (demand). The result for several years was a balanced world price of about $90.00 dollars per barrel. This was the price overlap high enough to continue total extraction but low enough for the market to bear ... for several more years as American consumers went deeper into debt. The world market price may never go much higher because there is diminishing wealth available for purchase. In theory, the price of energy can't decline much lower either for any length of time because it costs more and more to extract increasingly difficult or non-conventional oil. And, there are those who choose to keep the oil in the ground for future generations and wait for higher prices. Only a small minority who have access to remaining wealth will eat, keep warm, travel, and attempt to perpetuate the lifestyle of the past surplus-energy age … as long as their paper wealth and non-energetic assets can be exchanged for "real" energy.

Further infusions of paper wealth, financial bail-outs, and stimulus packages based on the hope of growth-revival in the future can not work because future energy growth, dependent primarily on oil, is no longer possible. There will be local bonanzas and temporary remissions because of new energy finds, new extraction technology, decreased consumption ("demand-destruction"), or improved efficiency, but the long-term trend must lead to system contraction because of the decreasing and dominate contribution from finite fossil fuels, specifically oil.

Suddenly in the summer of 2014, the Saudis upset this percarious equilibrium by refusing to reduce their market share for the benefit of new, more expensive non-conventional sources. The resumption of low-cost oil combined with the growing resistance to $90 dollar per barrel oil, caused a sudden glut in the oil markets including gasoline specifically for U.S, the world's number one consumer bloc. (As clarified in Chapter 1). The result was an immediate fifty-percent plummet of price to the mid-$45 per barrel range which climbed back to around $50 per barrel by the spring of 2015, but then dropped below $40 per barrel by the end of 2015. This is a good example of the impossibility to store significant amounts of energy. The world's tankers and storage facilities can only store a few days worth of oil waiting for the price to go back up while the futures markets have a field day as speculators try to out-gamble each other as to what will happen next.

As a consumer, you would expect to fill your gasoline tank to take advantage of the bargain price, but how many gallons and miles-worth can you hoard … on-board?

Back to the Energy Barrel

At some time in the past, present, or future, all energy (except nuclear or tidal) came, comes from, or will come to the earth from the sun. In our short, fossil-energy age, CONTEMPORARY-SUSTAINABLE inputs account for less than ten-percent of our total energy input and consumption. **Because of the dilute, sporadic nature of incoming, annual solar energy in its common forms of direct solar, wind,**

bio-energy, or hydro power, "all of the above" (from the sun) can NEVER remotely approach the amazing quantitative level of conveniently stored fossil fuels, which represent millions of years of prehistoric solar energy. Other non-fossil inputs like tidal or geothermal are tiny, site-specific, and poor in energy return. They cannot amount to more than diversions from our main focus, the imminent decline of TEMPORARY-FINITE-PRE STORED energy inputs.

Referring again to Figure 11, the INTERNAL SUBSYSTEM-TRANSFER of energy between inclusive subsystems is another source of confusion. For smaller internal subsystems like nations or regional societies, this is a most important factor. "We want what you got." Whenever a closed group of bio-species exceeds or decreases its carrying capacity (in its most basic form, food), or expands through population growth and the over-powering urge to survive (the "Selfish Gene"); territorial energy transfer is the ubiquitous normal result. Sometimes the transfer from one subsystem to another is hidden as a seemingly benevolent interaction. **But the effect on the total system-energy balance is net-neutral, like robbing Peter to pay Paul.** INTERNAL SUBSYSTEM-TRANSFER is simply a contraction of one subsystem to support growth in another. Externally-sourced TEMPORARY-FINITE fossil fuels, like oil moved from Canada and Mexico to another subsystem like the U.S. are examples of this category. They are only transferred from one region to another with no effect or difference in the larger global closed system.

The INTERNAL SUBSYSTEM-TRANSFER category also holds true for non-energy resources, but these are inert raw materials, clean water, or arable land **which cannot be accessed (mined), transported, or processed into useful goods and growth without ... energy**. Open the history book to any page, and this last category underlies and explains much of the glory and tragedy of our past. Resource wars and the slave trade are examples. Unfortunately, as surplus fossil energy winds down, we are beginning to regress to past human-nature. **We will do whatever it takes to maintain our personal subsystem's status quo by keeping energy inputs maximized, so we can maintain a high level of outputs, and at the same time, avoid subsystem contraction and ultimate collapse**.

For a completely closed finite system, like our planet, INTERNAL SUBSYSTEM-TRANSFER inputs are meaningless. We can't raid the moon for energy except for a little tidal gravity. We can access a small amount of internal geothermal energy, but it is site specific and can't contribute more than a tiny fraction of our needs. **Our only sustainable energy input source always has been, still is, and will be ... the sun.**

Specifics (energy outflows)

To postpone closed-system contraction (or "collapse") as TEMPORARY-FINITE inputs begin a precipitous decline in the post-fossil fuel age, our only legitimate option is to reduce (ration) the output losses just as with the water barrel or storage battery. We must try to control the leaks from the energy barrel.

ENERGY HEAT LOSS (entropy) from a warmer body to colder surroundings is a fundamental law of thermodynamics. On the earth macro-basis, there is no way to reduce this energy loss. Natural or industrial greenhouse gases only increase our planet's temperature a few degrees to a new equilibrium level. The resultant global warming does not reduce the overall quantity of energy output flowing back into cold outer space. However, the desired temperature of small sub-systems like living bodies or houses can be maintained with far less input energy by reducing micro-system HEAT LOSS through better clothes or insulation. This is the basis for efficiency savings and makes sense as long as decreased energy needs are not offset by continued overall system growth such as an increase in population. The common phenomenon of **increased efficiency leading to greater consumption** was clarified by W.S. Jevons at the beginning of the fossil-fueled industrial age in the 19th century, and bears his name, "Jevons Parodox."

ENERGY HEAT LOSS is a significant human concern because it is the ubiquitous energy output as we, individually and collectively, dissipate the heat from metabolized food or heated buildings. Higher food consumption for increased population results in proportionally more energy lost forever as heat and/or non-growth work.

SECURITY-DEFENSE energy output can be most important in any subsystem energy balance. Considerable precious energy may be expended by an individual, village, or nation. In some subsystems where energy inputs are copious, as in the fossil-fuel age, this is probably not a critical problem. But in other times, the energy cost to resist nefarious encroachments from other subsystems was often ruinous.

In future low-energy times, SECURITY-DEFENSE energy consumption and output will become more critical. An individual animal can exhaust itself by fighting for survival or to protect territory and food sources. The end result is the same: Sub-system energy contracts towards death. A nation subject to invasion like the Roman Empire can fail because of inadequate internal energy left after losses for SECURITY-DEFENSE. Should the last internal energy go to feeding people or defending them?

Another significant and often overlooked energy loss is the ENERGY INVESTMENT required (recycled) to acquire, and becomes included in, the input energy. The ratio of the two is named by the acronym EROEI (Energy Returned On Energy Invested). Sometimes this term is shortened to EROI and is usually analyzed for a specific energy source as the quantity acquired divided by the quantity expended for its aquisition.

When the ratio is greater than unity the acquisition is considered positive and the effort is energy-legitimate. The original extraction of conventional oil put this ratio over 100:1 but it is steadily declining for all fossil fuels as we deplete the easiest sources of prehistoric stored finite energy.

One example of a positive EROEI is wood. Otherwise our ancestors would not have survived. A few days with an axe could yield thousands of BTU's for warmth and the construction of shelter (system growth). In the fossil fuel age, it takes only a fraction of energy returned (invested as fossil fuel input in a chain saw or skidder) to harvest a much larger multiple of bio-energy. However, this loop is clearly unsustainable when more wood is harvested in one year (about one cord or 3000 pounds, wet, per acre) than slowly grows through the photosynthesis of incoming solar energy in the same period. In addition, soils, water, and minerals as ashes from the burned wood must be available and, in the case of nutrients, returned to where the wood is grown. Over-harvesting of wood beyond a steady-state sustainable level historically supported the overshoot of population and subsequent collapse of civilizations like Easter Island.

There are many other examples of marginal EROEIs, which contribute directly to contraction and decay. Confusion creeps in when other **subsystem** inputs or losses are combined with those used directly in an analysis. For instance, there is much argument concerning the EROEI of corn ethanol for fuel. Should all the inputs to grow the corn like irrigation energy, farm labor, fertilizer, fossil energy to manufacture farm equipment, and many more be included in the inputs? Should the heat from burning, rather than being returned to the source-land, the co-products (stalks, leaves, and ashes), be considered output? Depending on the methodology used**, the EROEI for corn to fuel-ethanol ranges from less than one to 1.5.** In addition, it is clear from our energy-balance model that the output energy for ethanol as fuel is finally lost as NON-GROWTH work rather than assimilated as INTERNAL GROWTH. **Using 30% of our corn for fuel instead of food for survival does not make sense from any quantitative or ethical view.** Yet, here we are. Drive up to the gas pump and fill up with 10% ethanol, a tragic testimonial to our short-sighted intelligence.

On a personal level, the ENERGY INVESTMENT to acquire new energy is different from NON-GROWTH WORK. Instead of going from point A to B and back again on the bike or a hike, we could expend the same food energy by growing more food or cutting firewood. Most energy loss is from energy consumers who do not work directly to facilitate energy inputs (including food from a farmer) but still use the energy. This has always been true of children, the aged, or infirm, but it is equally true for anyone not directly involved in the energy chain. **For instance, a productive adult who is suddenly out of work represents a continuing energy-system drain because he/she still has to eat, but contributes nothing to the input of energy into the system.** So, the system contracts.

This analysis could be carried further by questioning the energy requirements of (in no particular order) financial services, advertising, marketing, insurance, recreation, entertainment, management, building construction (except to save energy), space exploration, and so on. It may sound heretical to question traditional careers, but without copious surplus energy, any activity not directly related to essential transportation, food, or warmth will be greatly challenged.

All of the above occupations grew far out of proportion to basic farming. They flourished and grew only because of the abundant fossil energy age. Even essential needs like medical care and our judicial/penal system will be unsustainable as energy becomes scarce and expensive.

The effect of population growth

If there is no change in the energy inflow vs. outflow balance in our energy barrel analogy, there will still be a per capita lowering of energy level as long as the population increases. There will be less energy to accommodate each of more people. This could be simulated by imagining the barrel becoming wider in diameter.

THE CAR IS THE CULPRIT

This leaves NON-GROWTH WORK as the huge energy loss we could quickly control as if by turning a valve or fixing a leak. **If we were intelligent, we would immediately and equitably reduce vast quantities of wasted energy output by nation-wide coupon rationing of gasoline.** A close look at the energy flow model shows the TEMPORARY-FINITE flow of input (oil) energy matched by a massive gusher of outflow oil used for travel in all types of vehicles. As long as we had a fire hose filling our energy-system barrel we could tolerate a gaping hole in the bottom.

The age of easy travel is quite simply, the oil age. The two concepts are absolutely intertwined and inseparable because of the huge amounts of conveniently concentrated energy in a liquid fuel that can be carried along for the ride in a modern vehicle with an internal combustion engine.

As explained in Appendix 2, the mechanical definition of energy is the capacity, of some thing or substance to do work, where work is simply a force (to overcome drag or friction) while covering a distance. Distance is the most important term here because it implies movement or travel. If we use our personally stored energy (from food) for the work required to crawl, walk, bike, or run from point A to point B, at least we have an obvious result, we're in a different place B, unless we had just traveled back to point A. Our energy (to do the work) consumption is immediately apparent even if we only traveled around in circles wasting the energy through indiscriminate travel going nowhere.

Now, at the midpoint of the brief, profligate fossil-fuel energy age, we enjoy the luxury of riding in a 4000-pound chariot, cruising at seventy-miles an hour, enjoying the scenery, occupant warmth, and entertainment. Up to the last months of 2014, American drivers were burning gasoline at the rate of nine million barrels (almost four-hundred million gallons) … per day! On a personal basis, we use far more energy to travel a few miles to the supermarket than is in the food we bring home to feed a family of four for a week. In our overall energy balance, the EROEI in this case would be less than one which further contributes to system decay. In any other time in history, or in poorer parts of the world, such luxury would be a dream. There is no question that the vast American working and living lifestyle has evolved around the automobile as explained in Part I.

By 2005, oil and gasoline prices climbed to record highs as extraction and availability of inexpensive oil and related petro-products began to level off. The world entered a terminal recession with extreme geopolitical tension. Our future-growth-dependent-economic system ground to a screeching halt. Debt-based growth and expansion based on a continued energy surplus can no longer continue.

Other forms of energy cannot begin to substitute for the peaking production and consumption of oil. Yet our dependency and love affair with our cars continues unabated. As long as market forces control the price of gasoline, even if oil prices remain high because of production shortfalls all over the world, the wealthy can still out-bid the poor. But high gasoline consumption also affects the price of other related energy-intensive needs like food, diesel, and jet fuel. By 2014, there began a temporary surplus and price reduction called "demand destruction." Short-term price swings are exacerbated by speculative machinations of the commodity markets, which, in turn, lead to turmoil and further curtailment of world oil production whenever the price drops below profitable levels.

Why ration gasoline?

As long as market and speculative forces establish the price of oil while world extraction levels and then declines, we will continue to see erratic spikes. There are always fewer consumers who can bid the price back up. Price may not go much lower either because more expensive, non-conventional sources are financial losses. The various scenarios in Figure 2 predict just how quickly this end game will play out. It is certain that there will be less gasoline as oil extraction, the dominate component of TEMPORARY-FINITE input energy, inexorably declines. The tragedy of this inevitable scenario is that the love of the automobile trip, especially by those few who can still afford gasoline even at a higher price, will compete for the valuable energy required to produce essential food. Already this is happening as CONTEMPORARY-SUSTAINABLE ethanol and bio-diesel are substituted for oil.

High taxes on transportation fuels, as practiced for years in other industrialized countries, is a step in the right direction, but not much better than letting market forces balance price, supply, and demand. The wealthy can still outbid the poor. On the positive side, at least high fuel taxes encourage smaller cars, reduced travel by some, and alternative more efficient forms of travel. The price is somewhat stabilized, and a larger portion of fuel costs flows to the national treasuries.

Implementation of gasoline rationing

The devil is in the details, but in the electronic age national-coupon rationing could be done. No matter how inconvenient and unpopular, it is our only hope to quickly and significantly reduce energy output commensurate with the leveling and downward supply of all petro-fuels. We have no choice. We waste too much gasoline now for fast frivolous travel in huge vehicles. Our kids will wonder why we were in such a hurry to burn up the world's finite energy endowment and didn't save some for their survival. The present U.S. consumption of about two gallons per day per licensed driver should be halved to one gallon per day as soon as possible. **Perhaps a one-half gallon per day reduction would be an easier first step for several years. This would still allow for about fifty miles per day, per person, in an efficient car driven ... alone! The immediate reduction would be one hundred million gallons (2.5 million barrels) per day!** This is four times the 600,000 barrels per day from the Bakken "fracking" bonanza and almost half the total U.S. conventional oil extraction of five and one half million barrels per day. If we just slowed down, doubled up our passenger load, and drove smaller cars, we could achieve a reduction of one-half gallons per day, per driver **with no reduction in miles traveled, or impact on the economy.** This first step of gasoline rationing would ensure we all participate equitably.

The second step of rationing to one gallon per day for each of two-hundred million licensed drivers, would reduce oil consumption by almost five million barrels per day (1.73 billion barrels/year). This is ninety percent of total U.S. conventional oil extraction, just for one- half of our gasoline consumption! I challenge you to memorize and think about these facts. We could still each be driving twenty-five miles a day alone in an inefficient, 25 mpg vehicle.

To repeat, a central theme of this chapter (and entire book) is that rationing to one gallon per day of gasoline per licensed driver, instead of the two gallons per day average consumption, as it is now, would postpone over seventy-five percent of U.S. present extraction of conventional oil for future generations and the prosperity of our country. This is astounding! To put the numbers in perspective, the five million barrels saved each day is also approximately equivalent to:

- Twice the entire extraction rate of Iraq (after a cost of billions of dollars and thousands of lives.)
- About four times the projected output from Canadian tar sands oil without the environmental impact or the extremely high input energy of all types to mine, process, and ship the diluted bitumen.)
- More than half of the nine million barrel per day extraction rate of each of the world's largest producers, Russia and Saudi Arabia.
- The International Energy Agency (IEA) amazing prediction of a sudden doubling (in four years!) of US extraction rate to a level exceeding Russia and Saudi Arabia.

It could be argued that the same reduction in overall consumption might also be achieved if fuel efficiency was increased from 25 mpg to 50 mpg. True. But this would never happen unless **every** driver could afford to buy a very efficient tiny vehicle and drive it accordingly. There would be no universal incentive to do this and Jevon's paradox (drive and consume more because it is more efficient and less costly) would rule. Also, the steady increase of licensed drivers (about two million per year) would offset the decrease of overall consumption from rationing. **A fifty percent**

decrease in gasoline consumption can only happen with nation-wide, legislated, participation by all drivers. Never before in peacetime have we so desperately needed this profound level of understanding, acceptance, and leadership.

Other advantages of gasoline rationing

- Immediately, the cost of gasoline would decline and remain predictably low because of the reduction in U.S. demand. Although the price of oil reflects world wide supply and demand, our domestic consumption is so huge that a one gallon per day reduction in the U.S. would take one-sixteenth of world demand off the market.
- The lower per-gallon cost **plus** the approximate eight million dollars/per day saved by not being spent on gasoline would be a massive jump-start for all other non-travel sectors of the economy. This would be just the opposite of the steady economic drain of the past few years as Americans have steadfastly resisted curtailing their love affair with large fast cars.
- Overall health would improve as Americans get out of their cars and walk or bike.
- Electric transportation of all types would be encouraged. Mass transit would be favored and small electric vehicles would become popular as long as the battery problem is resolved.
- Safety would be greatly improved because of slower speeds and fewer vehicles. Small, efficient cars would be much more safe and appealing because of not having to share the road with huge, speeding vehicles.
- CO_2 emissions would be curtailed.

Problems

Of course there will be many questions if we use gasoline rationing to mitigate the decline of oil.

- With only gasoline rationing, there will be increased demand for diesel. This loophole should be left open as it would take years to substantially convert the U.S. passenger car fleet. By then, the reduced availability and increased price of world oil will reduce the overall consumption of all liquid fuels including other sectors of consumption, construction, heavy transport, public transportation, commercial, municipal, and farming.

The premium for distillate fuels like diesel, kerosene, heating oil and jet fuel will continue to increase and effectively curtail their use as well. Those passenger-car owners who must drive long distances have the option of converting to diesel which will further reduce gasoline demand. It is my personal opinion that European-style small turbo-diesel (TDI) passenger cars are less expensive, simpler, and a better value at 40 to 50 mpg than electric or electric-hybrid vehicles. Any increased emissions from the scandalized VW emission testing are more than offset by their inherently better efficiency.

- Farming needs for gasoline could also be handled on a case by case basis depending on farm income and tax reports. The same procedure could be an alternative for other commercial, legitimate businesses including taxis that don't convert to diesel.
- Air travel will slowly grind down if fuel costs escalate, providing the traveler can get to the airport on non-rationed public transport. It will remain an alternative for the wealthy, business, or emergency traveler.
- Ultimately, every commercial activity dependent in any way on automobile travel will be in jeopardy. But why should they be allowed to inordinately hasten the end of the oil age? They are products of the oil age and should reflect their dependency on the world's most critical finite resource. In reality, mitigation of the end of the oil age with coupon rationing will help perpetuate business-as-usual in changing times. Besides if traveling somewhere else for recreation is that important, the creative and affluent public will find a way (see heading "Implementation" below).
- The immense cost and complexity of coupon rationing are necessary evils if we are to navigate the end of the oil age without abrupt and total chaos. A small fee at issuance, like one-dollar per card for twenty million cards per day, would be enough to pay for the necessary infrastructure and personnel. Gas stations would be restricted to sale only to unused coupons or portions left, similar to a Walmart charge card.

Implementation

Each licensed driver who owns at least one registered vehicle would receive, monthly, from the state motor vehicle department (DMV), a book of coupons (plastic swipe cards) called Tradable Fuel Coupons (TFC's) or something similar. There could be one coupon for every ten gallons. These must be presented (swiped) at the gas station along with the payment for gasoline before pumping. The number of cars registered beyond one by individual drivers has no bearing on coupons issued. This way, very efficient vehicles could be used wherever possible, but not necessarily for car-pooling, family trips, RV's, or vacations. Surplus coupons could be saved or purchased on the open (tradable) coupon market.

Each retail gasoline station would have a small electronic machine to record usage and ascertain the unused gallons available prior to pumping. Like a phone card, the same machine could count the total coupons used and reconcile the number with the gallons pumped for each day. The TFC's would be similar to a pre-purchased gift card. They are in effect, a negotiable instrument like a ten-dollar bill. There is no record of ownership after initial distribution by the DMV.

Obvious exceptions to rationing would be government, municipal, emergency, security, and essential needs. In short, gasoline rationing would only effect private consumption. City folks who don't use much gasoline could sell coupons to those who need to drive long distances.

A network of private-enterprise clearing houses would immediately appear to function as intermediaries similar to dealers for gold and silver coins. The hassel will cause some consternation but reflects the true energy cost and threat to our future survival from profligate gasoline consumption.

SUMMARY

1. American gasoline consumption is the largest, quickest, and easiest candidate for controlled downsizing of oil consumption in order to mitigate and delay the inevitable post-fossil-fuel crash.
2. Rationing is the best, most equitable way to reduce consumption ahead of the inevitable decline in world oil production. Rationing would help reduce wealth disparity, keep price lower and more stable, as well as minimizing wild market-price swings. Now, uncontrolled demand, production costs, alternative liquid fuels (some competing with food), speculation, world tension, and steadily declining oil fields all interact. We are totally unprepared and out of control as we enter a new era of diminishing energy from fossil fuels. As the most intelligent species we should recognize the facts and plan accordingly.
3. Since American drivers use about one-eighth of world petroleum just for gasoline, a positive action to downsize will send a huge message to the rest of the industrialized world.

4. National TFC rationing would provide the time and price stability for exploration and development of alternative energy sources. The wild price swings we now have discourage long-range planning.
5. The street value of TFC s would have the added advantage of putting instant cash in the hands of poor and/or frugal drivers. This would redistribute real wealth from those who can afford and choose to consume more, to those who do not.
6. The equitable sharing and reduction of gasoline consumption would benefit all energy consumers, rich or poor. Rationing is the only way for peaceful co-existence when diminishing essential resources face steadily increasing demand.
7. Considering the dire challenges we face as we enter the second half of the oil age, it is certain that some form of rationing will also have to expand to other critical oil uses like aviation fuel, commercial diesel, and industrial transportation fuel. If we ration gasoline now, that eventuality can be postponed.
8. The controlled reduction of discretionary gasoline consumption will leave future petroleum feed-stocks available at a more stable price for the development of renewable fuels, heating oil, plastics, lubricants, agriculture, wood harvesting, strategic material mining, and a thousand other things that are oil-dependent, ubiquitous, and now taken for granted.

CONCLUSIONS

A thesis is offered based on the following logic path:

1. Energy is required for the growth or movement of anything of substance.
2. Our debt-based financial system of principal plus future interest is an example of growth-dependency which can only work when there is a surplus and commensurate increase in availability of energy.
3. Presently, over ninety percent of our energy is derived from TEMPORARY-FINITE sources led by oil at thirty-seven percent. The overall world-wide oil extraction rate of conventional oil has peaked due to the natural constraints of our finite planet plus increased costs and difficulty of extraction. There are many who argue that peaking has not yet happened, but if that was true, we still have only a few more years to prepare and take advantage of the extra time.
4. Suggested alternatives for oil are minuscule, site specific, delusional, or limited by annual incoming solar energy. Finite fossil fuels also represent solar

energy but only after the concentration of hundreds of millions of years in the making into three convenient forms. These are the laws of physics and math. No amount of wishful thinking or research grants can change them.

5. In order to keep our energy-intensive society from overshoot and then collapse from reduced real-energy availability, we have only one immediate option, to reduce real-energy consumption (loss).
6. The most wasteful energy loss (one-eighth of world oil) is the American use of gasoline for fast travel in large cars.
7. We could make a giant first step in oil reduction and mitigation of the imminent real energy decline by rationing the availability of gasoline equally between the wealthy and poor. One gallon per day per licensed driver would halve our U.S. consumption with little decrease in miles traveled if we drove smaller cars more slowly.
8. Electronic tradable coupons (TFCs) could be saved or openly traded on the open market. This would immediately transfer wealth from those who can afford to extravagantly use tomorrow's energy today, to those who choose to conserve and soften the impact of the imminent post-oil age.

Please give these thoughts a chance. Let them sink in. Test them against other proposals and pass them along to others, especially to those in a decision-making capacity. Our best hope is rapid, exponential, diffusion of ideas and information now possible in the electronic age. Without your personal energy, nothing will happen.

SUPPORTING MATH AND SIMPLE PHYSICS

Today, in very rounded numbers, two-hundred million licensed American drivers consume, each day, four-hundred million gallons of gasoline (ten million 42-gallon barrels). This gives every driver an average of two gallons or at twenty-five miles per gallon, fifty miles of travel ... each day! This quantity represents one-eighth of world oil production just for American gasoline. Does this unique level of gross energy consumption justify the premature demise of our energy-intensive civilization? **A better understanding of the following science will help reduce personal gasoline consumption even if rationing is not implemented:**

There are a number of ways overall consumption could decline, initially, with no reduction in distance traveled. For instance, let's look at the simple physics.

Starting back with the basic definition of work as expanded in Appendix B:

$$W = F \times D$$

Work (W) output is equal to the fuel-energy (used in the engine, minus heat losses) to provide the force (F) required to move the car, times the distance (D) traveled. **We can easily reduce the left-side (fuel required) to provide the necessary work (W) without reducing distance (D) by decreasing only the force (F) on the right side. This would initially leave the distance traveled (D) unchanged.**

When driving **on the level**, (F) consists of only two significant drag forces that require work (as fuel-energy): rolling resistence (R) and air turbulence (T). (T) might also be called wind resistance or air drag. (T) and (R) represent energy lost as heat, forever, from pushing air molecules aside as the car passes through the air or flexing the tire as it rolls over a surface.

In algebraic terms: $F = R + T$

If we take a minute to further contemplate these two terms, R and T, it will clear up much confusion, bogus information, and save much fuel.

Rolling resistence (R) is simply a numerical coefficient for a type of wheel (on a specific surface) times the weight on the wheel. For a hard rubber tire on pavement it is about 0.015 times the weight on each wheel. If tire pressure is increased, this coefficient is less. If vehicle weight is less, rolling resistance (R) is also reduced. This is why bike tires are pumped up to 100 psi, and bikes work best on hard pavement. Lighter vehicles use less fuel than heavier. That's all there is to it. Quantitatively, the rolling resistance of a car is not very much, about 45 pounds for a 3000 pound car. A human can push a car, albeit slowly, **on hard level pavement.** There is not much fuel/energy that can be saved by using a lighter vehicle or pumping up the tires for a few percent reduction of (R).

Air drag (T for turbulence) is a little more complicated but needs to be clearly understood as we move into a low-energy future. Air drag (wind resistance) is especially important in electric cars which have much less onboard energy. Air turbulence (T) only begins to be a factor above about 20 mph and is where and why we can make significant reductions of work (fuel) required. **We will rarely go fast after the oil age is over.**

There are three important factors multiplied together which make up the air drag (T). At very high altitudes, thinner air density also reduces T, but we will leave that advantage for airplanes out of our analysis. The total equation for air drag is:

$$T = A \times S \times V^2$$

The frontal area of a vehicle is (A). Simply put, a vehicle with twice the frontal area will have twice the air drag. This is why airplanes are packaged like a cigar and travel in thinner air, and why bicyclists crouch down to go fast.

Yet, we insist on driving cars (trucks) as big as barn doors because fuel has been so plentiful and inexpensive for the last 100 years. Smaller cars would immediately save a lot of work (fuel) with no change in speed or distance traveled.

The second term (S) is a factor which is determined by the shape of the vehicle and the way air (wind) flows around it. Intuitively we know that a streamlined egg shape will move through the air with less turbulence than a rectangular brick, yet we still waste gas driving cars (trucks) shaped like one. A modern streamlined car has a shape factor (known as the coefficient of aerodynamic drag) of about 0.3 and there is not much more we can do about this factor except drive well-shaped cars. In addition, both frontal area and shape factor are not significant **below speeds of about 30 mph because the total air turbulence (T) is small and less than rolling drag (R).** We are not concerned about shape and frontal area while pushing a baby carriage or riding a lawn mower, but we certainly become aware when riding a road bike.

This brings us to the third factor, speed (velocity, V). This is the most important of all three terms because it is squared. If we go twice as fast (churning and heating air) it becomes four times as restrictive. (Two squared equals two times two equals four.) Do we really need to drive fast to get somewhere a little sooner? Fuel rationing would immediately encourage slower driving because speed will use up the coupons quicker over a shorter distance. President Carter was right with the 55 mph speed limit in the 1970s as the U.S. went over the peak of domestic extraction. If we go twice as fast, we require four times as much work, but we get there in half the time so the total energy consumed over a distance is four times one-half, or twice as much. **To drive eighty miles in one hour theoretically takes twice as much gasoline as taking two hours at half the speed. Are we Americans really so important we can justify using our future fuel just to get to our destinations sooner?** (This explanation is only basic physics and is not exactly true because the engine and drive train energy losses are only existent for one half the time. Higher gear ratios also reduce these internal losses so they are not proportionally higher.

The above basic analysis does not account for changing velocity or climbing hills. **If we use the car's brakes to slow the car or control speed on a descent, the energy loss from heat is considerable and lost forever with nothing to be gained but worn-out brakes.** This is why electric cars use regenerative braking to return some of the precious energy to the batteries. The answer is to think of the car as a pendulum, slowing as much as possible to let momentum crest the hill and pick up speed as much as possible to avoid braking at the bottom. Accelerate very slowly and anticipate stops way in advance so rolling resistance and wind drag substitute for heat loss from braking. Slower speeds and "hyper-driving" techniques are not always compatible with today's traffic, but this is the context for considerable fuel saving without having to reduce distance traveled. If you understand all this theory, go out on the interstate and try it today. **See how quickly you will be aggressively tailgated and passed by huge vehicles in a great hurry to get somewhere else. Only equitable coupon rationing will slow traffic, save great quantities of fuel, and make for safer, more relaxing travel.**

Reducing any combination of the three terms (A, S, V) offers the best possibility for considerable immediate fuel savings with no initial reduction in distance traveled or change in current lifestyle.

Remember, the fuel (energy) conserved is far more critical for other sectors of our energy system including food, heating oil, social services, commercial transport, national defense, petroleum-based products, and all the other ubiquitous uses of petroleum as a food or feedstock. Converting coal and natural gas to automotive fuel, converting to diesel passenger cars, or plugging in hybrids only avoids the issue by substituting other TEMPORARY-FINITE energy inputs and moves us farther down ... the wrong road. Rationing would dictate changes in driving habits equitably between rich and poor. No other concept would yield as much significant energy reduction for our future survival; yet it needs to be legislated so all share the effort. We need to get started now to conserve our nation's remaining oil endowment for the coming oil crash.

CHAPTER 8

Food-Energy: the Fragile Link Between Resources and Population

Every successful species harbors a genetic drive to reproduce more numbers than can be supported by a stable, sustainable environment. The limitations are food, non-energetic resources necessary as building blocks for life, competing species (also hungry for food), a hospitable ambience, and adequate physical space. Those who best adapt evolve as conditions crowd out or supercede those that don't. **This is simple Darwinism and most often infers a short, harsh individual life of competition and survival.** Changes in the environment by natural causes and/or environmental destruction add additional challenges to the status quo, and favor those species which are fortunate enough to adapt, **or smart enough to plan ahead.**

THE FOOD-ENERGY BALANCE

Humans acquire their energy from food which at one time came directly (or indirectly farther down the food chain from other plant eaters) from plant photosynthesis of incoming solar energy. As quickly as new offspring begin to grow on their own, increased demands are placed on the food supply although at first it may be as food-energy still supplied by a supportive parent(s). **As long as food is available the population will increase to the limits of the species' range, individual or collective skill (including tools), and individual energy available to procure the food and/or avoid being food for others**. It should be obvious that population is therefore limited first and primarily by the ability to access and store food. A quick look back to the energy-barrel analogy in Figure 11 may help explain the potential for growth or contraction of any system (including a human body) based on the input-output balance of energy, either stored from the past or acquired contemporarily, **but cannot be borrowed from the future.**

To continue our analysis, we will focus on the food-energy balance required for human survival **without** relying on a temporary surge of non-renewable energy capital, e.g., fossil fuels, or imported food from another location or time (inferring a surplus somewhere else). **For hundreds of thousands of years our predecessors lived as hunter-gatherers in a precarious, but on average, ratio which had to be larger than one between food-energy returned on (personal) energy invested (FEROPEI).**

Cultivation and agriculture

Then, about ten thousand years ago, humans learned to utilize favorable, unique, local growing conditions and crops. The age of agriculture began. A slight improvement of human FEROPEI, combined with reduced energy requirements and risk related to excessive travel, provided a tiny energy surplus to support the beginning of recorded history. The extra energy made possible early civilizations and the construction of ancient monuments many of which have survived to this day. To build anything of substance, the energy has to come from the excess over and above the primary requisite-energy required for personal food and survival. **Instead of a hunter-gatherer barely able to procure food for himself and enough progeny to perpetuate the species, a hard-working farmer with favorable ecological conditions could now feed additional dependents plus non-farmers.** Energy-intensive travel was still limited by human personal mobility, draft animals, and wind power for sailing.

Because of agriculture, the average FEROPEI improved, possibly up to 6:1. This provided the steady-state support for long-term societies like the Chinese, **but could not support continued growth** of extensive, non-agrarian expansion like the Roman Empire. Many societies flourished, then collapsed because of the inevitable conflict between growing population and the limitations and degradation of local food-carrying capacity.

Additional energy

Following the collapse of the Roman Empire, slow adaptation to wind, water, and draft-animals gradually improved local agricultural output by reducing the direct dependency of food output from human-energy alone. The food-energy return on human labor input (FEROPEI) slowly increased up to the 10:1 or 12:1 range. Still, population was held in check by disease, poor health care, unpredictable diet, infant mortality, and short life spans. But, the new energy surplus made possible multiple layers of non-farm population, societal admininstration, and cities. Marketing, money, and laws evolved on the backs of peasant (or slave) farmers. There was enough extra energy and manpower to support armies and wars required to wage territorial, resource, and cultural disputes.

A very erratic and slow increase in world population continued up to just several hundred years ago. Additional food sources and room for growth came primarily from exploration and settlement of new lands. Malthus's prediction of population limits was temporarily proven wrong because the exploitable world seemed limitless. Then, suddenly, inventions of new ways to exploit the convenient vast stores of finite fossil energy, far beyond renewable wind and water, made possible and began the industrial age. As would be expected, the unprecedented utilization of non-human energy escalated the food-energy available for human consumption and population growth.

A new (very short) high-energy age

The utilization of vast stores of pre-stored fossil energy, beginning with coal two-hundred years ago, and followed by oil and natural gas, suddenly jumped the ratio of food energy returned on personal energy invested (FEROPEI) to as high as 300:1. One farmer could now feed three-hundred people instead of six by himself or twelve with the help of animal power. Concurrently, in the same short period, as would be expected, world population soared from one-billion to seven-billion. This was due to many factors directly related to the sudden energy bonanza. Access to formerly remote lands, genetic crop improvements, inorganic nitrogen fertilizer, pesticides, energy-intensive farm equipment, irrigation, refrigeration, packaging, and long-distance mobility all contributed to the modern lifestyle now enjoyed by the industrialized world. Concurrently, giant strides in medicine and health care vastly increased life span and population. **But still, all must eat with the same basic individual requirement of 8000 BTU per day.** Figures 4 and 7 show the direct correlation between the sudden increase of oil consumption and population. Figures 1 and 8 combine both as per capita utilization by the average world inhabitant. Figure 1 shows the huge disparity in oil consumption between the world average and extreme consumers as are living in the U.S.

In the past several decades the "green revolution" maximized food production and made possible the seven-billion humans now needing food. However, this final push is unsustainable and is causing new problems; for instance, genetically modified (GM) foods may be linked to new health risks.

Resilience, an advantage of crop diversification, is absent. New pesticides, monoculture, and herbicides also lead to super-bugs and environmental contamination. The societal improvements promised by capitalism and industrialized agriculture are, in effect, just more examples of temporarily polarizing wealth between the masses and the few who control the system. **Finally, and obviously, the mechanization of modern agriculture cannot continue without oil.**

The link must break

Now, after a one lifetime span of almost free energy and resultant copious food, the entire world faces the imminent decline (and eventual demise) of finite, fossil-fuel capital and therefore must inevitably face a return to the food-production default range with a FEROPEI as low as 6:1 or, at most, 12:1. This assumes individual farmers can retain a semblance of traditional agriculture, knowledge, hard work, and renewable energy, while drastically reducing non-food energy expenditures for travel and keeping warm. This is the "end point" we must return to in the next 60 to 80 years while, **in the same one-lifetime span**, reducing the numbers to be fed from the present seven billion back to, at most, one or two billion who **must also be located very close to their food source.** Without fossil fuels, food can no longer be produced in one area and shipped thousands of miles to market. Nutrients must also be returned to their source. **To suggest that the world will be able to feed the UN projected population of nine billion by 2050 is totally incomprehensible in the face of declining oil. This food-energy disconnect is shown as a "food gap" in Figure 4 and this book's cover.**

Can we return to "sustainability"?

Homo Sapiens will survive. Our ancient ancestors lived off the land and survived ice ages (without metal or plastics). But, in order to avoid total collapse first, we must clearly recognize our predicament in quantitative terms and define exactly what to do. We must get started immediately to allow time for a commensurate population reduction through natural attrition instead of famine, war, and disease. **We will not make it without three basic prerequisites necessary in the time and direction available:**

1. **Explicit knowledge and broad publicity of what to do, and why we (including you) must get started immediately, especially networking this story.**
2. **Negative population growth at a level of not more than one child per female (or male).**
3. **A systematic reduction of American per capita energy consumption from 22 b/b/y to 3 b/b/y, including rationing, in the next thirty years.**

All must be done.

We should avoid wasting and/or fighting over the remaining oil. Never before in recorded history has there been so singular a resource as oil for food, or the urgency for a controlled descent from the ephemeral peak of energy usage we enjoy today. **No other species has achieved the feat of anticipating and systematically executing an energy and directly-dependant population reduction.**

In my office piled high with pertinent web print-outs I have one, in particular, a comprehensive classic (Theoildrum.com/node4628, Oct 20, 2008). It is a paper by Peter Salonius, a Canadian soil microbiologist. The title, *Agriculture: Unsustainable resource depletion Began 10,000 Years Ago*, and six parts including Part 4, *Intensive crop cultures are unsustainable*, cover the entire theme that "human population numbers will have to be brought into balance with the sustainable productivity levels of the local ecosystems upon which they rely for their sustenance."

In Part 6 is a wonderful concluding paragraph:

> *Balancing of human numbers to the productivity of their supporting local ecosystems may be accomplished by planned attrition, much lower birth rates and the economic dislocations and hardships that a retreat from classical economic growth will incur, or the balancing of human numbers may be accomplished by a catastrophic collapse imposed by natural resource scarcity. The species with the large brain must make the choice between economic hardship and catastrophic collapse.*

PERSONAL FOOD PRODUCTION

At this point, I will switch from a macro-overview of fossil energy and the energy-food dilemma to a few comments on all aspects of small-scale farming. Those who grow at least some of their own food will be far more prepared for the coming changes in the food-energy paradigm. **"When the music stops, know where your chair is."**

As a context for this part of the discussion I offer a brief background: I grew up during and after the second world war on a small "gentleman" farm in western Massachusetts. We were surrounded by commercial dairy farms, tractors, and work-horses. We had several riding horses and did our own haying. During the summers before going off to engineering college in 1952, and with the help of a prewar 9N Ford-Ferguson tractor, I would grow several acres of heirloom, Golden Bantam sweet corn "picked while you wait" to ensure the sweetness we take for granted today with sugar-enhanced hybrids. Fast forwarding to 1980, I left the fast paced business world. My wife, our one-year old son, and I moved to rural Maine to pursue new interests, a more peaceful life than world-wide new-product development, and the possibilities of self-sufficiency on an old 175 acre farm.

With nine generations of "hard-rock" Vermont farmers in my genetic makeup, I always feel a strong sense of satisfaction and renewal while growing our own food. I will also add, in the last thirty years, and with the focus on our own well-being and food sources, we have gradually segued from a home-grown meat diet to nearly one-hundred percent vegan. We attribute much of our personal good health to this transition. Now, after becoming immersed for the last decade in the looming oil-energy crisis, I am especially concerned about if, and how, we could feed just my wife and I without oil. We can argue about when, but someday within the several decades, oil and the plentiful super-market food we take for granted will be in short supply and/or very expensive. A gallon of gasoline in my chain saw or sixty-year old John Deere already seems extremely important. Long distance bananas, pineapples, and avocados will no longer be staples in our diet.

We must all start immediately to grow as much of our own food as possible. This is the fun part and is the subject of a vast popular movement highlighted by innumerable books, magazines, and web sites. Square-foot gardening, raised beds, and permaculture are the new rage. Everyone should begin some form of personal food independence. We don't need thirty-million acres of lawns. Flowers aren't very filling either. An added bonus is that at least a small part of our personal human-energy, inherently imbedded in our bodies for food acquisition, will again be put to use as nature intended ... accessing more food. We will be far healthier and become more insulated from the poor nutrition and fragility of a giant, long-distance, energy-intensive food chain. In addition, if we were smart and resolved the battery problem, we could begin to integrate new solar-electric technology into our more self-sufficient low-energy lifestyle.

Without fossil fuels it will be impossible for the vast majority to live in cities or urban centers that are not directly surrounded or down river from extremely productive farmland. We do have some hope for a future far better than our hard-working ancestors who had no electricity or photovoltaic "slaves" ready to go to work when the sun shines. The magazine, *Acres*, (acresusa.com) carries a vast catalogue of hundreds of books from organic farming to eco-gardening. Another unique publication that seriously addresses post-oil age food challenges is *Countryside and Small Stock Journal* (countrysidemag.com). See quote on my book's back cover by *Countryside* founder: J. D. Belanger.

In addition to growing our own food, and while energy is still plentiful, we should begin to accumulate a personal and family food reserve. Dried, canned, or bottled foods will keep for years. Our rural ancestors always kept a winter's supply as well as seeds and feed for their animals. In addition to traditional canning and a root cellar, we can have a modern electric freezer to extend a surplus harvest, as long as grid or PV electricity is available. We have been routinely testing and eating dried lentils, split peas, rice, and beans that have been sealed and stored for over fifteen years. They may not taste quite the same as fresh, but when boiled are perfectly safe and would be far better than starving.

How much do we need?

From a base-line, human-energy requirement (assuming 5,000 btu per pound energy-equivalence of **dry** weight for food) and an average of 8000 btu (2000 kilo-calories) per day, **we each need about 500 pounds of food per year. Remember, we are always talking about dry weight.** Most of our food is harvested or purchased with over 60 or 70% water content (like our bodies) and water does not provide any of the energy/fuel we need. Fat and meat have higher concentrations of energy, closer to 10,000 btu per pound (pure fat is like petroleum products at 15,000 btu per pound). For food independence and security for one year, a vegetarian family of four would need a stash of a ton (2000 pounds) of dry weight plant foods.

Current "just-in-time" super market inventories will only last a couple of days in an energy crisis. The traditional dry-storage of dependable foods like grains, beans, and corn (and a heavier weight of potatoes, squash, root crops, or fruit) now make sense. If we don't eat them, we can replant or barter with neighbors. As an alternative, for ready-prepared survival food, good for up to 30 years of storage, a good place to start is: www.thefoodstore.com.

More quantitative details

To grow your own food, think of green leaves as photosynthesizing solar panels which chemically use solar energy to combine water and nutrients from the earth with CO_2 from the air. Because incoming solar radiation energy is very dilute and intermittent, it requires considerable time and area to grow a pound of dry weight (not including water) biomass. This is why the solid parts of a plant, like the stem, tree trunk, gourd or seeds, need much more green foliage spread out onto a wide area like a solar panel. It's the same energy storage problem again. The solid part of the plant conveniently concentrates and stores substantial energy for use later when needed for survival and reproduction.

To supply the average, individual, annual energy-requirement of 500 pounds (dry weight) in a one year growing season, about 1/4 of an acre (11,000 sq. ft.) of good productive garden surface area is required. Vertical plants like corn and pole beans need less area because they access more sunlight energy for a given ground area, but need more soil-nutrients and water. In addition (usually overlooked or ignored fact), if this system is to be sustainable, the human waste or equivalent biomass and nutrients removed **must be returned to their original source.**

Any permanent removal or burning of biomass destroys the cycle. This fact, plus extremely poor EROEI, are among the many reasons why biofuels for motive power are absolutely non-sustainable.

In addition to space requirements and nutrient-import/export problems, to grow meaningful quantities of food, we must confront the human-energy input requirement. **Assuming near-by arable land is available (not more than walking distance), how can one strong farmer do all the ground preparation, planting, cultivation, and harvesting to feed himself and five others from 1½ acres at the historical maximum of a 6:1 FEROPEI?** How do those, who are not farmers, reimburse the farmers? They must have something as valuable as food. How do we power the farm from a personal to nationwide level? These are the basic quantitative, physical, economic, and ecological limitations we must respect. They are the foundation for our entire food discussion, from a personal to a national level. To start, we could become much more food-efficient with respect to waste. But that would only postpone the inevitable tension between mouths to feed and supply limitations.

Richard Heinburg in his book, *Peak Everything*, suggests we need "fifty-million farmers" in a U.S. without oil and 300+ million people. This number agrees exactly with my 6:1 FEROPEI ratio. In their book, *A Nation of Farmers* Astyk and Newton would rather see everyone growing food. For protein, the affluent western world has become very dependent on animal products including dairy. Unfortunately this is a double-edged sword. Domestic animal food requires approximately ten times as much energy input for the net output of human-food energy. In addition, this practice has made us less healthy. Animal protein may be (along with excessive sugar consumption)one of the primary reasons Americans suffer from obesity and have over twice the per capita health care costs as the rest of the world. Studies have shown that many industrialized-world health problems can be traced directly to the substitution of animal protein for vegetable protein. See: Campbell, *The China Study* (Benbella Books, 2005); Lyman, *Mad Cowboy* (Touchstone Books,1998); Rifkin, *Beyond Beef* (Penguin Books, 1992); and, from an extreme athlete's perspective: Jurek, *Eat & Run* (Houghton Mifflin, 2012).

Water

So far in this short discussion about food and agriculture we have not considered the obvious need for plentiful water at the right time and place. **Droughts and climate change are hastening the tipping point of our demise.** The summer of 2014 was unprecedented for high temperatures and lack of water for more than 50% of the U.S. **California is nearing the point where rationing will be needed for equitable distribution.** Irrigation depends on power and an adequate water table. When there are food shortages we forget that, historically, population tends to increase to the limits of carrying capacity during the good years. Then the weather is blamed for crop failure and catastrophe; not the extra population added during the good times. The increased tension ("link") between a growing population and declining crops invariably leads to societal collapse. Which was the root cause, over-population or climate change?

Input power and energy

To provide a self-sufficient, personal level of food production using any size of garden larger than a few square feet **without** fossil energy is a huge challenge. Draft animals are problematic because of their requirements for year-round input energy (food), animal husbandry, replacement, and specific horse-drawn equipment. Besides, they don't interface well with electric power like lights or motors, especially in winter.

A modern answer to provide agricultural energy as a supplement to human labor is the solar panel. The farmer has complete control of this source and may or may not be connected to an electrical grid. Just two 175-watt PV panels in the direct spring and summer sun are equivalent in power to five adults working non-stop **without rest or food-input energy.** Four of these panels would provide one horsepower. They could provide the same energy/work in one hour as a draft horse, again, without food or rest.

The problem, as always, is to store the energy in a sufficient quantity to continue doing serious work for an extended period. A farmer can't graze a battery-powered tractor. He must return to, or have on-board, solar panels and plentiful sunshine. **He must also have provision for battery recycling as discussed in Chapter 5.**

A solar-powered vehicle like the 48-volt golf cart with two panels equaling 350 watts of PV described in Chapter 5 can do it all except the initial plowing or harrowing. With a long 14-gage extension cord, the on-board 2500-watt inverter will easily power a 24" rototiller/cultivator. It can also be used for personal transport of up to

100 miles with a 150 ampere hour lead-acid battery pack and no solar input. It can power a 3½ hp electric chain saw (try that with a workhorse), water pump, or an electric lawn mower if still needed after most of the lawns have been converted to food production. The solar-powered golf cart concept/size **is an excellent portable power source which could also supply grid back-up or minimal alternative power for one family's residential needs.** For gardening needs, one golf cart could support a neighborhood of up to six families and provide the mobility and food-transport between market and farm at fifteen mph. Traditional draft-animal power is limited to five mph and is much less desirable on a personal basis because of the needed skills, animal husbandry, and hay and grain input energy to feed the animals (horses or oxen).

Access and ownership

Still another hurdle barring the return to personal agriculture is who owns the land? In the days of share cropping and tenant farmers, a portion of each individual farmer's meager output answered that question in a form of rent. **Without ownership, legal access, fuel for travel, and nearby homes, how can individuals who are not close to, or do not own, arable land become farmers? Unless these questions are answered long before we run short of oil there will be a total collapse of property rights, law, and civility.** We cannot expect millions of small farmers to be suddenly, evenly and peacefully, dispersed across the land, or to invest their precious energy into reclamation and production of land they don't own.

By now it should be clear why excess energy for any non-food related form of travel, frivolous or otherwise, will no longer be available. **As this reality sinks in, we see why extending the oil age as far as possible with reduced consumption, improved efficiency, and equitable fuel rationing; all combined with controlled population reduction, are our best and only options to mitigate the descent from the oil age.** Familiar long-distance diesel for boat, rail, and air travel, and transport also cannot continue without liquid petroleum-based fuel. Chapter 7 expands a detailed discussion of energy, transportation, and rationing. All these perplexing challenges lead back to **the central theme beginning in Part I of this book: our only possible path to avoid collapse of our modern lifestyle is to immediately reduce American per capita oil consumption from the present level of 22 barrels per person per year (b/p/y) to the world average of 3 b/p/y.**

Local (community) food production

Contiuing the argument for rapid growth in personal agriculture, the aggregation of like-minded individual families into sustainable support groups is gaining popularity throughout the industrialized world. Community Supported Agriculture (CSA's), localization, eco-villages, and transition towns are becoming popular movements. These all offer potential economies of scale and specialization that are superior to individual farming. There is also more resilience against weather variations, individual farmer's health, mistakes, pests, and poor soil management. If all goes well, there should be a surplus for local sale but commercialization adds the requirements for distribution, product quality, dependability, marketing, and some form of currency beyond barter.

On the negative side, community-sized food operations are still limited by the same space vs. population-to-be-fed constraints as the individual family farms that are its base. No longer will there be availability of food imports from other communities or more distant farms. The limitations of decreasing fossil fuels, especially oil, will be universal. There will be less transportation energy to package, preserve, and ship food as we do today.

There will be less possibility to recycle exported nutrients. Ultimately, every farmer faces the physical need for at least 1½ acres of nearby arable land and provide one hard-working individual to support five people beyond him/herself. Community centers (towns) will have to revert back to the pre-oil days with not more than a 15 mile walking radius to the surrounding farms. This circular area of over 450,000 acres is more than enough for wood lots, pastures, and recreational space for thousands of families, providing the minimal area of tillable land is adequate to feed everyone in the encompassed community. The wood lots will also have to provide for home and public building heating at a sustainable level not exceeding one cord per acre per year. **How would the wood be harvested, fitted for firewood, delivered and reimbursed?** Even one cord per year is not truly sustainable if the nutrients (ashes) are not returned to the wood lot.

A solar-powered golf cart-sized machine cannot provide the power and energy storage for ground breaking like plowing, rototilling, harrowing, and other high-power tasks we take for granted. A scaled-up answer, other than reverting to draft animals, could be a larger solar- powered tractor as described in Chapter 5. A 20- to 30-horsepower tractor can do these heavy tasks as well as hay-mowing and bailing. Unfortunately, 1200 pounds of lead-acid batteries will only store about 12 kWh of energy at 60 % depth of discharge (DOD), the equivalent of 1½ gallons of gasoline.

This is only enough for up to one hour of typical 15hp work, enough to prepare about one-quarter of an acre. But it is far better than doing it by hand or having to support draft animals on a year-round basis.

More reality

A nearby 1½ kW (eight-panel) array requires at least eight hours of direct sunlight to recharge the large tractor. The 1200 pound lead-acid battery pack would cost about $3,000 and have to be recycled (where?) every 5 years or 500 cycles. An alternative might be a $20,000 lithium-ion battery pack which could store four times as much energy and plow more than one and a quarter acres (like a good team of horses in a long day); but the 1½ kW PV array would then take at least four days to replenish the 48kWh of energy. Grid charging works well and the 120 volt dc tractor battery pack matches the 120 volt ac grid voltage for charging with a simple rectifier and no transformer.

And always we must ask **the question: where does the grid energy come from?** Coal, natural gas, nuclear? All are finite sources and wind or hydro power are not even close to supplying a fraction of our total energy needs. We can hope that community-scaled agriculture has a potential yield and resilience beyond personal farming, but closer analysis reveals the same limitations plus additional problems of labor management (shirkers?) and equitable distribution of food, income, and nutrient return. More on these subjects in Chapter 9.

National (U.S.) Food Production

Unfortunately, without liquid fossil fuels, neither personal nor community-scale agriculture could supply a tiny fraction of our national food system and the population we have today. We cannot suddenly transition from one million farmers to fifty million. The present system of industrialized agriculture is based on only twenty-percent of the farms to supply eighty-percent of the food. **The U.S. agribusiness system and nation's population are totally dependent on oil (and natural gas for nitrogen fertilizer) for daily food requirements.** Without oil, one farmer cannot possibly feed 300 people, especially if the consumers are 2000 miles away. We can dream all we want about personal and community food production, but the numbers are totally unrealistic on a national scale. Maybe 100-hp electric tractors are conceivable, but where is the time and capital to build these machines and solar energy to recharge the batteries going to come from? **This is where reality sets in and why we must immediately begin to ration liquid fuels to buy time,**

conserve our country's oil, and shift away from an unsustainable system in the same (less than one lifetime) time frame required to downsize population. It seems totally absurd and myopic for Americans to go on burning through one-fourth of the world's (including our) remaining oil, to continue our oil party as fast as we can, when we can clearly see that, in the lifetimes of most of us alive, we will be short of food. A frivolous trip today wastes finite fuel better saved for a combine (or chain saw) thirty years from now. **We cannot plan on biofuels because they compete with food crops and take about as much input fossil-fuel energy to grow and process as they yield for useful work. Biofuels are also totally unsustainable because they abruptly break the soil-energy-nutrient cycle.**

We can envisage vast teams of mules or horses on the high plains, but we would still need concentrated energy for food preservation, packaging, and long-distance transport. As oil becomes more expensive, we are now leveraging it further to grow even more food with larger tractors and less labor. Presently, it is possible for two farmers to plant one-thousand acres (!) in one day with a 36-row corn planter. Americans can still go to the supermarket and buy pork or chicken "specials" for 99 cents per pound. It is exactly this trend and dependency on petroleum fuels that has driven the small farmer out of competition and idled millions of acres of old marginal farm land closer to the consumer. This is another example of profit-motivated capitalism leading us farther and farther out on the finite-energy limb without the foresight to turn around and climb down. Also, a steady increase in fossil-energy-dependent food-cost, combined with decreased surplus available for export is having devastating effects on third-world consumers who became "hooked" on imports, overpopulated, and lost their local subsistence food system. The question asked many times begs an answer: "Are humans smarter than yeast?"

A sovereign nation like the U.S., with a remaining endowment of oil, could conceivably extend an oil-based, high-energy, nation-wide food system, but only for a few more years. **Remember, fact: one-half of all the oil extracted and consumed to date, in the history of the oil age, has been used in the last 25 year-long, one generation span.** As is well documented, the extraction of United States (including Alaska) oil extraction peaked 40 years ago, then declined to one-half that rate. Now, because of higher oil prices which supported improvements in technology and production from non-conventional sources we have returned to 75% of U.S. peak extraction rate of the seventies.

The 2012 election year platforms from both parties promised to restore growth and prosperity while ignoring the steady addition of over three million mouths (approximately one-percent growth) to feed each year. Smaller, high-energy countries like

Japan and the UK are in much worse shape. Meanwhile, we Americans continue our lifestyle of fast travel in monster vehicles, fueled, in part, with corn-based ethanol. Very few Americans are ready to hear about downsizing population and rationing gasoline. The price of gasoline reached almost $4.00 per gallon in the spring of 2013 then collapsed in mid 2014 to half that level due to a temporary glut from decreased demand. How many Americans see fuel prices as part of a national food problem that can not be solved with proposals to substitute natural gas, ethanol, or algae biofuels for oil?

Global food production

Figure 4 shows the two-lifetime correlation (link) between oil extraction and population as a basic theme for this book. We are pushing the limits of our finite global carrying capacity made possible with one critical resource, oil. **One local region or community might temporarily flourish while others collapse, but the planet is a closed system with nowhere else for humans to go, and only "finite" resources to access.** All subsets of this complete system must average out to the numbers and X-axis time scale in Figure 4. Even in the oil age, already a third of the world is suffering from food shortages. Anarchy and riots will become more frequent as in pre-industrial-age revolutions when food ran short. The "food gap" is growing. We are at a point in time where fewer individual nations have a chance to cope with their own food security, energy conservation, and population reduction with a minimum of human suffering. Instead, we Americans expend prodigious amounts of our children's energy-legacy to police the troubled world and keep the remaining oil and exported food flowing into wealthier countries like ours. It is impossible to talk about the future of food without including the subjects of population and geopolitics. **We must begin a conversation about food and peak oil and how this nexus ("perfect storm") relates to almost every other serious subject in the daily news. Climate Change (man-made or not), water supplies, top soil loss, desertification, fisheries depletion, habitat destruction, peak phosphorous, peak potassium, and chemical pollution are all part of a bigger picture and cannot be considered separately.**

CHAPTER 9

Downsizing and Localization

LIMITS AND LOCALIZATION

Continuing beyond Chapter 8, and at the risk of redundancy, I will add a few more thoughts to these two directly-related terms. The word "limit" is fundamental to our human predicament. There are a number of books in the included bibliography which use this word in the title. It is difficult to comprehend how, with finite (synonymous with limited) energy and resources, a system can continue to grow. A farmer cannot keep adding livestock and feed more people when his farm is limited in size and ultimately, in productivity. Continued attempts to exceed the physical limits with more difficult sources of fossil fuels and/or new technology will only postpone the day of reckoning and exacerbate the consequences while population continues to grow.

These are the same limits to the concept of "localization." A growing number of communities around the country consist of a handful of individuals who understand the unfolding energy crisis. The Transition Town movement is an example. **Maybe if a few of us just circle our wagons we can weather the storm and live happily in a self-sufficient community.**

Population

It's the same problem regardless of the community size. If a local group increases in numbers per the methodology defined in Chapter 6, and exceeds the carrying capacity ("limits") of its resource base, the per capita food supply will most certainly fail. Figures 4 and 7 show population with two children per female (2 cpf) continuing to grow regardless of community size and without immigration or exporting people to other closed communities. Growing the community's food requirements is the goal, but how about the canned goods, protein (animal or otherwise) condiments, bananas, oranges, chocolates, paper goods, matches, soaps and all the other extras we routinely pick up at the super market?

And remember, any draft animal assistance for work or travel must also be fed from the same working land which reduces the food available for human consumption. It takes about one-fourth of the productive farm area just to "power" the farm with horses or oxen. The same one-fourth rule of thumb holds true for biofuels grown on the farm assuming the extensive infrastructure and capital investment are available. Biodiesel or ethanol production are far beyond the skills and equipment available on individual farms.

In addition, **for long term sustainability, all nutrients and energy-equivalents of food that leave a specific area of arable land must be returned directly to their original source in the form of manure and waste products ... including "humanure."** (See *The Humanure Handbook* by Joseph Jenkins.) To export food or rely on imported fertilizer and compost defies the principals of closed-system self-sufficiency. Also, long-distance movement of food and/or return of equivalent nutrients is impossible without fossil fuel energy or water transport. The Romans were fortunate to live on the Mediterranean Sea, but still found that land transport was limited to a few hundred miles because the energy required to feed the draft animals for their muscle power was greater than that in the food being moved.

Local heat and hot water

Without finite fossil fuels, the only renewable source of domestic heat other than sporadic solar-thermal is firewood or some other form of biomass. It takes about one acre of good wood-lot to yield one cord (2000 pounds dry) per year without depleting the base forest. Without fossil fuels how will this wood be harvested and transported to the home except by muscle power? The only answer is solar-electric as described in Chapter 5. **In addition, repeated cycles of cutting firewood are no more sustainable than removing hay or food from the land without returning an equivalent quantity of the removed nutrients.**

Travel

A localized community is restricted by limited travel. Up to a 15 mile radius from the social center is all that is possible by foot, animal, or off-road bicycle. This was the case with small towns in pre-industrial days. Asphalt-paved roads will be crumbling. Maintenance will be curtailed as heavier oil byproducts of fuel production for all forms of liquid-fueled vehicles become unavailable. **We will never go far or fast again, or move large loads without oil or synthetic liquids (synfuels) from**

coal or natural gas ... both requiring massive amounts of energy input and contributing to climate change. Biofuels for travel or transport will not happen because of unavailable fossil fuel energy input. See Chapter 4.

Most people live too far from bodies of water for barges or sailing. That is why early population centers began near the potential for water transport or where canals were dug ... by muscle power. We've already used most of the high-energy anthracite coal that was necessary for trains. Wood power for steam engines, or gasification will be very limited by availability, the energy required to produce, and very poor (about ten percent) efficiency. Again, the only modern alternative is electric power which can access the energy from PV panels, wind, or hydro power, as supplied by a third rail or carried along in a heavy, expensive battery.

Everything else

For the needs of a self-sufficient "localized" community, we could start an alphabetical list and not get much farther than "b" or "c" before becoming hopelessly bogged down with all the day-to-day needs we take for granted in our high-tech, oil-fueled life. Where else will these be made, and with what energy, in a post-oil age? **It is well documented that far-simpler societies than ours collapsed because of specialized interwoven dependency.** See *The collapse of Complex Societies* by Joseph Taintor. Back to the "b's": bulbs, bullets, batteries (of all types), bottles, bicycles, sanding belts, v-belts, bolts (and nuts), brushes..and so on.

And the "c's": computers (repair?), cans, chains, coins (will they suffice for barter and wealth indebtedness?), candles, canvas (no more plastic tarps or hoop-house covers without petroleum-based feed stocks), copper cartridges for lead bullets, cement, coffee, clocks, cloth, copper wire, cords, and many more. Steadily increasing complexity will soon cause the entire system to unravel. **No single community can be responsible for everything. Specialization creeps in until everyone is an integral part of a much-larger, very fragile whole. Many diverse hands, minds, and local materials are overwhelmed by the failure of any single cog, energy, or other essential non-renewable resources.**

In addition to the basic necessities of food, heat, and shelter, each functioning civil organization must provide the human energy for myriad other activities we take for granted; like education, health care, local administration, social interaction, formal interaction with other community centers, civil control, and security from a local to national level.

By now, the argument should be clear: a few cannot expect to isolate themselves from the highly-structured, high-tech life style we now enjoy. Despite the comradeship and mutual support found in a "localized" movement, there is no hope that local resilience or transition towns will shield us from the imminent macro-energy and over-population crises. Concerned people (You?) that begin and participate in local groups are usually those who have the best grasp of the enormous challenges we face. You should be the most vociferous by "networking" these thoughts on the largest scale possible. Finally, a functioning local community cannot expect to "export" excess progeny to other localized communities which are also struggling to keep within the limits of their own finite carrying capacity.

PART IV

TWO DIRECTLY RELATED SUBJECTS

To conclude this book we turn our attention to:

How the fossil-energy age will end while hidden behind a confusing interaction of economics, terminated-growth, and the cost of oil.

What kind of leadership can best lead us into an acceptable post-hydrocarbon age? Traditional democratic forms of government will not function for growing masses of unhappy constituents each trying to preserve his or her piece of the comfortable past, or just trying to survive.

CHAPTER 10

Economics in an Energy-Constrained Future

ENERGY, THE LIFEBLOOD OF AN ECONOMY

Economic success, growth, and an affluent (happy) consumer lifestyle depend directly on an abundance of inexpensive readily-available energy. Conversely, the quantity and type of energy can have a very adverse effect on the surrounding environment and world ecological balance. It then follows that leadership and politics, for the governance of civilized societies, should be intimately concerned with the tight connection between economics and energy. Now, at this unique and critical time in history, we are facing the unprecedented terminal decline of oil, our prime energy source. Following soon, in the next several decades, will be diminishing availability of all finite fuels.

The advocates of related subjects, for instance climate change (man-made or not) or stimulus proposals for continued economic growth, do not factor in the difficult, if not impossible, transition and immense challenges facing us as we enter the second half of the short fossil-energy age. **Without energy to make things happen, nothing grows, moves to a new place, or expands.** Bodies wither and die, civilizations contract and collapse. Yet there are leaders and experts who would lead us to believe otherwise, that "finite" does not mean the dictionary definition, or a shortfall will magically produce "substitutes." Oil still supplies over 37% of our total energy including 90% used for all modes of modern travel and most of the fuel for construction and transport. In addition, we've come to depend on thousands of petroleum-based products from lubricants to plastics. There may be "plenty left," but oil is steadily harder to find and more expensive in terms of input energy and wealth required for extraction from a growing percentage of unconventional sources.

Equating economics, energy, and oil

At this point, I will insert a stand-alone essay I wrote seven years ago. It was posted on Theoildrum.com/node5621 on August 1, 2009 and led to a long, wandering 79 page blog- discussion about the validity of electric tractors, and the potential of nuclear energy without the support of petroleum inputs. This is another example of seemingly infinite web information falling on deaf ears and not making a speck of difference as we (the U.S. and world) slide off the per capita oil cliff. **I don't profess to be an economist, but it certainly appears that most economists don't want to confront energy, finite resources, and population growth. Politicians surround themselves with the most renowned, comforting economists and perpetuate the standard myths that scarcity will produce substitutes and increased cost will drive perpetual availability.** Anyway, following is a humble retired-engineer's opinion:

A. NOTHING OF SUBSTANCE MOVES OR GROWS WITHOUT ENERGY

This includes a body, a bird's nest, population, a building, a road, or a civilization. Energy is arguably the most important word in the dictionary. Oil is presently the world's primary source of energy, providing almost 40% of all energy and over 90% of transportation fuel. (Fuel is another term for energy.) Energy is necessary for and can be represented by warmth or heat resulting in a higher temperature over ambient surroundings. Most of the world's energy came from or is coming via radiation from the sun's fusion, albeit dilute and sporadic as it reaches the earth. Exceptions are nuclear fission, geothermal, and tidal.

Power **IS NOT** energy. Power is only a measure of the rate that energy is being used or changed into a different form. It **IS NOT** synonymous with energy yet is loosely used that way by the media and "experts" which further confuses the public. Energy **CANNOT BE BORROWED** from the future. Next week's food won't assuage today's hunger.

B. WORLD PRODUCTION OF PRE-STORED OIL (representing millions of years of conveniently stored solar energy and photosynthesis) **HAS PEAKED.**

Conventional (light and easily accessible) oil reached a maximum of over 75 million barrels per day in 2005. All liquid fuels including tar-sand oil, heavy oil, deep off shore oil, polar oil, natural gas liquids, and bio fuels peaked at about 85 million barrels per day in the third quarter of 2008 (see Figure 3 for 2014 update). These numbers are historic facts as presented by the International Energy Agency and our own Department of Energy (eia.gov). U.S. production peaked in 1972 exactly as predicted in the 1950's by M.K. Hubbert. This fact resulted in the "energy crisis" of the seventies and a sharp increase in the price of oil as well as a temporary reduction of world oil production. This early warning was quickly forgotten and superceded by vast new sources of world oil from our Arctic, the North Sea, Russia, Mexico, South America, Africa, and the Mid-East.

C. WE LIVE IN AN ECONOMIC SYSTEM ENTIRELY DEPENDENT ON GROWTH

Our prosperity needs the promise of a future return of principal **PLUS** interest to justify the investment of present principal. This worked well for the last one-hundred years as long as there was always an excess of cheap pre-stored fossil energy available to "fuel" the growth. (For this premise, we will ignore inflation and speak in terms of real growth.)

D. THE CRUX: NOW THAT PRE STORED ENERGY, REPRESENTED BY OIL, HAS PEAKED AND IS IN TERMINAL DECLINE, GROWTH AND OUR ECONOMIC SYSTEM CANNOT CONTINUE. (Conventional oil is still "plateaued" into 2015, but per capita oil has steadily declined. See Chapter 1.)

Prosperity, food to feed a growing population, an oil-based transportation system, and new building are all forms of energy dependence, which must now go into terminal decline.

This is a geophysical constraint, not choice or something that can be avoided by changing the laws of physics, political action, increasing demand, or wishful thinking. Civilization and our cheap-energy lifestyle are on the verge of collapse. The longer we deny the situation and try to perpetuate the party, the more severe will be the crash and fewer will be our options.

E. ONE SOURCE OF CONFUSION IS THE HIDDEN PRICE OF OIL

If oil is becoming scarce, why is oil (sometimes) less expensive? This is where things become more complicated. The price of oil only reflects the delicate balance of multiple transactions between consumers and oil producers. If consumers have a declining ability to pay from past, stored wealth then there is less real value to support ever-increasing costs to extract the remaining more-expensive oil. As more of the world is producing less of everything (especially energy-dependent food) because of energy-curtailed growth, only the decreasing sources of cheap oil are competitive. Out-of-work consumers cannot support new oil exploration and the remaining expensive, non-conventional sources, which were supposed to save us. So, the price deflates to a lower level.

If the economy begins to revive a little bit, the increased demand drives the price of oil back up until the declining, remaining wealth cannot support more-marginal, more-expensive sources. Fewer, poorer customers result in more-desperate suppliers, the only ones who can still produce relatively cheap oil, or who must keep their population under control at any cost. The end result is the beginning of the second half of the 200 year oil age. The first half (hardly more than one lifetime) was typified by growth, prosperity, and increased population. The second half will only be the opposite unless we recognize the enormity of our dilemma and quickly initiate emergency damage control and drastic measures such as are summarized in the acronym: LEARN … Localization, Education, Adaptation of solar power (in its several varied forms), Rationing (of remaining fossil energy starting with gasoline), and Negative population growth (on our terms rather than waiting for more abhorrent catastrophes). Nowhere in this essay have the terms "global warming," "climate change," or "environmentalism" been mentioned. These are obviously related to energy, the hyper-consumptive fossil fuel age, and are of dire concern. It is this writer's opinion, however, that these issues tend to divert focus from the imminent energy-economic crisis, which is not well understood and conspicuously absent and avoided in the media.

ECONOMICS, FAST FORWARD TO 2015

Now, almost ten years since "peak oil" became a controversial term and the great recession of 2008 is behind us, the world and U.S. economies seem trapped in a web of contradiction. In six months, the price of oil suddenly plunged below the fifty-dollar range. The stock market is bouncing around an all-time high. Unemployment is down below seven-percent. **The economists, politicians, and media are self-reinforcing the conventional wisdom that demand and new technology has (and always will) provided the necessary oil for the resumption of perpetual growth. Any talk of population control is contrary to the economic wisdom that: "new**

growth will be supported by new consumers." The words "finite" or "contraction" along with climate change are relegated to a fringe of abstract intellectuals.

Meanwhile our leaders lurch from debt ceiling, to sequester, to threat of shutdown as they fail to admit or understand why the growth of the first half of the oil age cannot be extrapolated onward and upward. It is fact that very wealthy individuals, banks, and corporations have cornered most of the trillions of dollars of U.S. residual wealth. They are reluctant to jump into new capital investment because they seem to intuitively know that largely underpaid or unemployed masses can not support continued growth. The middle class economy has been decimated by the export of good-paying manufacturing jobs along with increased costs for food. An ever-increasing population doesn't get paid enough to be significant consumers. Increased numbers of older generations must keep working because of zero interest return on their investments. They compete with new, younger job seekers for depressed wages which are not enough to cover the cost of living. Meanwhile, the wealthy own the remaining assets, have a substantial income, entitlements, or have a financial interest in the temporary resurgence in domestic energy. They have few places to park their inordinate share of the economy to protect it from inflation. We need bold new leadership that understands the limits of growth and reacts to an informed constituency that demands action.

It's up to you to get involved. Raise your voice. Listen carefully for substantive discussion of declining energy and/or increasing population (including immigration) in the 2016 election year. Which prospective candidate could become a leader as discussed in the next chapter?

CHAPTER 11

Leadership and Politics, How Will We Get There?

POLITICS AND GOVERNMENT

Regardless of their form of government, the great civilizations of the past, like Mesopotamia, Mycenaean Greece, the Roman Empire, and the Chacoan Society in our own desert southwest; grew and ultimately fell because of the tension between population, climate, and dependency on sporadic, daily solar-energy input. Even slaves, who were the preferred source of work for the affluent before the industrial age, required food/energy input. The underlying need for energy is universal for every successful species. Humans are no exception. **Yet, we take for granted the easy life we've enjoyed in the industrial age because we learned how to enslave millions of years of concentrated ancient sunlight-energy in the form of conveniently-stored, finite, fossil fuels.** Read the 2012 title that eloquently addresses this subject; *The Energy of Slaves: Oil and the New Servitude* by A. Nikiforuk.

In America we are governed by the constitutional framework of a Democratic Republic. We elect our lawmakers and leaders for a system of laws, checks, and balances. We have a federalist concept of shared rights between states and centralized government. This arrangement, despite the difficulty of long-distance travel, worked admirably well for two-hundred years. **Our population was sparse and expanded into a land of seemingly unlimited natural resources.** A second resource bonanza, this time of pre-stored and nearly-free energy, fueled a booster shot for continued unfettered growth. This unprecedented surge of easy energy made possible a high-technology lifestyle, surplus food, freedom from drudgery, and magical travel. Capital investment, based on the promise (premise) of never-ending growth, and return of investment plus additional profit, provided the financial backbone to "capitalize" on the fossil fuels. **A common citizen could now live as a king in pre-industrial times with the "Energy of Slaves" at his/her beck and call.** Despite the setbacks of two world wars and one great depression, economic and population growth expanded in unison.

Then, just after the dawn of the twenty-first century, the abundant fossil-energy foundation for this unprecedented prosperity began to level off onto a bumpy plateau. By mid-2005, conventional crude oil, by far the best and only fuel for modern transportation and easy agriculture, quantitatively peaked in world production at just over 75 million barrels per day. As this is written ten years later, that "peak" is still being traversed, but not exceeded. These are unarguable, historical facts per the U.S. Department of Energy (DOE) regardless of political or media obfuscation. The resultant, inexorable tension between growing demand and constrained supply led to a sharp increase in the cost of energy and a monumental recession. The inevitable correction began in a housing market dependent on easy lending; both directly dependent on continued, extrapolated perpetual growth.

By summer of 2014, Americans were spending over one billion dollars per day of their dwindling income just for gasoline. Most of this to drive vehicles too big and too fast (see Chapters 1 and 7). This does not include fuel oil, diesel, and jet fuel.

The increasing bite into the family budget left less for mortgages, local and national discretionary spending, and especially food, which is directly tied to the cost of energy and therefore also becoming more expensive. A steady destruction of demand, beginning with gasoline in 2005, finally caught up with the production of all liquid fuels. The result led to a sudden over-supply of oil and the collapse of price going into 2015. The normal reaction in time of glut has been to store as much liquid fuel as possible. Since it is difficult to store energy, five-hundred million barrels in storage sounds like a great deal but is only twenty-five days of U.S. consumption. The numbers only serve to convince the public that we have copious energy for future use and any talk of "peak oil", rationing, or a terminal oil-age falls on deaf ears.

A tale of two freedoms

With that background, we can better understand much of the divisiveness that has invaded our two-party political system. Liberty and the pursuit of happiness for all are fundamental tenants of our original constitution. **When he was president, FDR's second Bill of Rights taught "Four Freedoms": freedom from want and fear** in addition to speech and religion. A conflict between the first two of these basic expectations and personal liberty now comes into sharper focus because the increasing scarcity of cheap ubiquitous **energy can no longer provide freedom from want for everyone.** Finite, natural limits cannot supply enough food and fuel for an ever-increasing population. This dilemma is already the norm in the third world and is steadily creeping up the income ladder in our industrialized societies. This fact underlies why we have gross wealth disparity with a very few rising above a shrinking middle class, in turn, absorbed by a rising tide of the poor.

As energy becomes less available and more expensive, should wealthy individuals have unfettered "liberty" to access fuel, food, and all other energy-dependant needs, even if it increases "fear and want" for others? This dichotomy must be addressed. It is becoming physically and mathematically impossible, even in the U.S., to feed, keep warm, and maintain mobility for the present population while inexpensive oil supplies are stretched to the limit and nearing the point of permanent decline. This concept may be difficult to accept, but it is very real for the 80% of the population, who have only 20% of the remaining wealth. **Without cheap energy we can no longer all be hyper-consuming Americans.** Those who still have the financial means can outbid those who do not. The wealthy naturally resist policies intended to share this wealth. **At the same time, the total number of consumers continues to increase while job growth has stagnated. The result is a soaring nineteen-trillion dollar national debt and no chance of satisfying entitlements without economic growth.**

The growing, underlying conflict between the "freedom of liberty" and "freedom from want" has directly infiltrated our politics and exacerbates tension between the right and left. **Long-term growth, jobs, prosperity, leisure pursuits, and all things dependent on plentiful energy can no longer continue for everyone**. There may be temporary remissions because of the temporary oil glut, improved energy-use efficiency, new extraction technologies, and continued borrowing of wealth from the future. But true long-term economic growth, in excess of inflation, can not be sustained without the underlying foundation and promise of plentiful, inexpensive fossil energy.

So far, neither political party will admit to permanent energy contraction. The conservative right promises renewed growth through decreased taxation on business, new innovation, and new investment. The liberal left promotes redistribution of waning wealth to the steadily-increasing masses, including immigrants, who are moving closer to missing the basic necessities. **Both sides advocate increased exploration, efficiency of use, and technical progress. Both sides ignore the geo-physical limitations of the short fossil energy age.** The right promises renewed growth from fossil fuels previously off-limits in parks, federal lands or off-shore preserves. The left defers to reduced consumption, infrastructure repair, and renewable alternatives as the answers. Either direction leads to the conflict between a stalled-out economy dependent on continued growth, and a growing populace, all needing employment, food, social services, and long-term entitlements.

Neither side can provide "freedom from want" to the majority. Our democratic system swings back and forth in each voting cycle from the incumbent party, which did not deliver, to the other side promising to do better, and a return to "the good old days." Reagan, Clinton, and George W. Bush were lucky to take charge when oil was resurgent and cheap. Carter became unpopular after one term when he was confronted with peaking U.S. oil and world oil price turmoil. Obama appears to be suffering the same growing discontent as Carter because his term coincides with a time-zone in history of maximum world oil production regardless if the oil comes from friendly or unfriendly sources. Yet, in the past year he is not credited with lower gasoline costs.

This brings us to the question of **which party or basic system of government can best handle the realities of contracting energy and expanding-population.** Is a democracy of the people, for the people, and by the people still viable or will anarchy rule? In a free election will an individual vote for personal gain and survival, or will he/she lean toward the common good of the populace? **On a personal micro-basis, would an empathetic human (or any species for that matter) go hungry and starve if necessary to feed as many as possible of his neighbors, if only for just a few more days ... after which they will all starve together?** These are questions and choices we must confront. Lack of awareness and/or continued inaction only diminishes our chances of, at least mitigating the same fate of previous crashed civilizations which did not respect the inevitable clash between increasing consumers and finite and/or contracting energy resources.

Deferring to Plato (427–347 B.C.)

Our struggle to find direction for a challenging future is not unique. As regional societies crashed as nature and numbers played out their conflict, at least one great mind pondered the subject of leadership for the benefit of future generations. In my opinion, the best model is found in the dialogue for an ideal state. Typical of Plato's search for the ideal general form, in "*The Republic,*" we hear him conversing with Glaucon as he suggests the ideal "philosopher-king":

> *Until Philosophers are kings, or the kings and princes of this world have the spirit and power of philosophy, and political greatness and wisdom meet in one, and those commoner natures who pursue either to the exclusion of the other are compelled to stand aside, cities will never have rest from their evils,–no, nor the human race, as I believe,–and then only will this our state have a possibility of life and behold the light of day.*

It sounds like Plato is speaking directly to us. Now we face the most serious challenge ever for our survival, for our country, and all of civilization. An elected leader dares not mention the two clashing elephants in the room, growing population vs.

declining resources, without fear of "political suicide." We can argue about which subject: population, energy, or environmental degradation, is most serious. All must be considered together as a "triple crisis." The response for all three is common and urgent. Returning to the need for profound leadership:

Shakleton, Churchill, Autocracies

There have been isolated exceptions of profound leadership when certain disaster loomed. A great example is the true story of Ernest Shakleton who saved all his crew after their ship, *Endurance,* was trapped and crushed in the ice of the Antarctic Ocean. Another is the leadership of Winston Churchill throughout the critical days of World War II. These are examples of when, in times of undeniable crisis, only profound and "wise" individual leadership, most often insisting on extreme sacrifice (like rationing), could suffice.

Another related example is the one-party political system of China. However unpopular, at least decisions are made regarding population control, alternative energy investment, and resource acquisition. This type of authoritarian leadership is better than uncontrolled, genetic human nature which, in times of stress, automatically defaults to a Darwinian survival mode of "take everything you can get and run." This "me vs. we" genetic drive combined with continued reproduction, by as many as can survive, is obviously more successful for long-term species survival regardless of the trauma it entails. Excellent references for these subjects are: *The Selfish Gene*, by R. Dawkins and, J.Bligh's *The Fatal Inheritance.*

My hope for a drastic course-correction, at this late date in the oil age, will require a grass-roots movement which, in turn, supports "wise philosopher-kings"; leaders who clearly understand the growing tension in an economic system based on continued growth and declining energy. **There still may be hope for perpetuation of a vastly downsized modern lifestyle, but only if we admit to the seriousness of our terminal illness and not be lulled by bogus panaceas and/or political promises necessary for election.** We are clearly at a tipping point. In the last eighty years (one lifetime) we have consumed approximately one-half of the world's original endowment of conventional and non-conventional oil. In the same period we have used a large percentage of high-energy coal, natural gas, and high-concentration, fissionable uranium. Together, these finite sources provide over 90% of today's world energy with the U.S. (with 5% of the world's population) consuming about 25% of the total.

In the process, we in the industrialized world have destroyed much of our environment, possibly beyond the point of no return. The next human lifetime, starting now, will be extremely challenging. The world economy is like a giant bus stalling on a hill. Experts are scurrying around trying to get the engine running again just as many more passengers climb aboard. Very few want to check the fuel tank. We're running low.

Nowhere in this chapter or book is there a suggestion of dictates, edicts, or mandates from the leadership. The only hope is that the "wise" leadership will educate the public so well that constituents will understand and demand the changes we must all make together. This mass movement would lead to legislation (like gasoline rationing) by a majority in congress. It is unlikely that birth control laws would ever be proposed (or enforced!), but the public should clearly understand the choice between suffering, competition, and starvation by many or, the alternative: an acceptable long life for fewer. There is a large percentage of Americans that avoid involvement by deferring to faith or higher power. This path circumvents the laws and logic of numbers and physics, and absolves believers from participating in urgent corrective actions mutually beneficial to all.

MORE RECENT PUBLICATIONS

There is a flurry of publications that wade directly into the growing contrast between demand and supply. Unfortunately, most authors perpetuate the prevailing myth that a "replacement" fertility rate of 2 CPF (child per female) will do the job. The subtitles speak for themselves:

Full Planet, Empty Plates: The New Geopolitics of Food Scarcity, Brown (Norton, 2012). This is the most recent of a long series by the now retired leader of Earth Policy Institute. Like the others before it, the focus is on climate change, but by segueing into "food scarcity" the underlying context of population vs. energy is no longer on the back burner.

Winner Take All: China's Race for Resources and What it Means to the World, Moyo (Basic Books, 2012). Dambisa Moyo is an economist and freelance writer with a Ph.D. in economics from Harvard and a Masters from Harvard. Her new book moved to number thirteen on the New York Times best sellers list and speaks quantitatively about resource shortfalls and China's drive to corner what's left through both monopolies and monopsonies (meaning cornering the markets by underpricing sales and overpricing buying throughout the world). Typical of her thesis, from page 174: "When it comes to food, water, energy, and materials, for example, there are clear signals today that these resources will not be enough to go around in the near future. As we witness the groundswell of the global population and as wealth and prosperity expand, global supply is struggling to keep up, but investment lags behind and nature has its limits."

Scarcity: Humanity's Final Chapter?, Clugston (self-published, 2012). Order through Chris Clugston's website: nnrscarcity.com. With a prophetic forward by William Catton, author of *Bottleneck* (2009) and *Overshoot* (1982), this amazing effort lists, in quantitative detail, the remaining quantities of **all NNR's** (Non-Renewable Natural Resources). If we don't run short of oil first, virtually every other essential component of our modern industrialized civilization will soon follow. Needless to say, Clugston does not hold out much hope for the future.

The Race for What's Left: The Global Scramble For The World's Last Resources, Klare (Metropolitan Books, 2012). Michael Klare is the author of fourteen books dealing directly with resource depletion and related geopolitics. His books define the ultimate clash between decreasing oil and increasing demand.

The Crash of 2016, Hartmann, (Hachette Book Group, 2013). More ominous words from the prolific writer who also wrote *The Last Hours of Ancient Sunlight.*

Any Way You Slice It: The Past Present and Future of Rationing, Cox (The New Press, 2013). A comprehensive treatise on the need for rationing when critical resources run short. Going into 2015 the "R" word is moving into the mainstream conversation as water becomes critical in California, another example of climate change being blamed when population and consumption push regional carrying capacity to the limit.

The Limits of Growth Revisited, Bardi (Springer Books in Energy, 2011). This is one of a flurry of recent publications updating the original *Limits of Growth* predictions. Ugo Bardi is an expert on energy and natural resources. He is also President of the Italian Section of ASPO. **His analysis is a complete, contemporary validation of the original forty-year old computer modeling and system dynamics** by Dennis and Donella Meadows, Jorgen Randers, and William Behrens III. It's amazing how accurate these original predictions were with regard to resources, population and pollution. For further reading, the 1976 book, *Strategy for Survival: an Exploration of the Limits to Further Population and Industrial Growth*, Boughey. (W.A. Benjamin, 1976) is an early and extremely comprehensive analysis of the "limits of Growth" work as well as other similar studies from that time.

Going Dark, McPherson (Publish America, 2013). For the extremists who argue that anthropogenic-caused climate change will be our demise, This author predicts "the near-term exit of *Homo sapiens* from this planet..by the2030's"

Debunking Economics: The Naked Emperor Dethroned, Keen, (Zed Books, 2012). This is a 450 page book by a professor of economics that delves into every dark corner and fallacy of conventional economics.

Supply Shock: Economic Growth at the Crossroads and the Steady State Solution, Czech, Steadystate.org (New Society Publishers, 2013).

Energy: Overdevelopment and the Delusion of Endless Growth, Butler, (Watershed Media, 2012).

Immoderate Greatness: Why Civilizations Fail, Ophuls (Createspace, 2012).

Technofix: Why Technology Won't Save Us or the Environment, Huesemann,(New Society Publishers, 2011).

Afterburn: Society Beyond Fossil Fuels, Heinburg, New Society Publishers, 2015. The latest by U.S.'s leading author and proponent regarding Peak Oil and the end of Growth. Page 104.

Not the Future we Ordered: The psychology of Peak Oil and the Myth of Internal Progress, Greer (Karnac Books 2013). This most prolific author has recently published a flurry of related books. Others are: *Collapse Now and Avoid the Rush, After Oil2; The Years of Crisis,* and *Decline and Fall: The End of Empire and the Future of Democracy.* All reinforce my personal thoughts in a much better writing style.

The Five Stages of Collapse: Survivors Toolkit, Orlov, (New Society Publishers 2013).

Too Much Magic: Wishful Thinking, Technology and the Fate of the Nation, Kuntsler (Atlantic Monthly Press 2012). From one of our earliest and best authors about peak oil and the aftermath.

Peeking at Peak Oil, Aleklett (Springer Science + Business Media 2012). A complete, in-depth look at all aspects of peak oil by a professor of physics at Uppsala University of Sweden. Kjell, Aleklett was a co-founder, with Colin Campbell, of the original international ASPO.

Dark Peak, Fehling (Founders House 2015). This is one of the best of a flurry of scary novels describing life after the oil age.

Peak Oil: Apocalyptic Environmentalism and Libertarian Political Culture, Snyder-Mayerson (University of Chicago Press 2015). This new title is an amazing, comprehensive review of the erratic uncoordinated peak oil movement. The author includes copious endnotes and results of his own questionaires.

Myths of the Oil Boom: American National Security in a Global Energy Market, Yetiv (Oxford University Press 2015). An in-depth review of the fallacies of resurgent American oil extraction.

The Energy World is Flat: Opportunities from the End of Peak Oil, Lacalle and Parrilla (Wiley 2015). This is a bonnanza for peak oil debunkers. To quote page 66: "The end of the oil age will not happen because we ran out of oil. And it will not be a sudden and terrible shock that will bring economic hardship to people. The end of the oil era will be gradual, cyclical, and will open a new and more prosperous era for humans." I could not disagree more.

Tumbling Tide: Population, Petroleum, and Systemic Collapse, Goodchild (Insomniac Press, Canada 2013). This is the other extreme from *The Energy World is Flat.* A doomer's delight.

American Theocracy: The Perils and Politics of Radical Religion, Oil, and Borrowed Money in the 21st Century, Phillips (Viking 2006). A comprehensive study of the history and synergisim between oil, religion, and finance in the U.S. today.

Lifeblood: Oil, Freedom, and the Forces of Capitalism, Huber (University of Minnesota 2013). An excellent textbook for all the subjects in the subtitle.

The Oil Age: Understanding the Past, Exploring the Future. This is a new journal published four times a year by the Petroleum Analysis Centre, Stabil Hill, Ballydehob, Co. Cork, Ireland.

Presidential election 2012

The Romney Energy Plan promised "An Achievable Goal: Energy Independence by 2020." This will be possible from "surging energy production, combined with the resources of America's neighbors, ...". The results: "The emergence of an Energy Superpower." **Quantitatively, for "oil independence" in eight years, the Romney agenda would have magically jumped U.S. oil production two-fold from the present eight million barrels per day (down from ten at the peak of U.S. extraction in 1970) to the present U.S. consumption of over sixteen million barrels per day.** Plus, we are not told that "other liquids" like natural gas liquids and bio fuels have been added to the mix to reach the present U.S. consumption of nineteen million barrels per day. Romney continued that "we must return to the glory days of Ronald Reagan" with no mention that Reagan's era coincided with the last world remission of cheap oil.

The happy right-wing promise of renewed and everlasting abundance is perpetuated by several other new books. Two that are getting much media coverage are: Gilder, *Wealth and Poverty*, and Forbes, *Freedom Manifesto: Why Free Markets Are Moral And Big Government Is Not.* Both of these authors hammer away at the same traditional economist themes:

- Government causes scarcity.
- A free market with unfettered technology and human ingenuity will always find new substitutes for any commodity in short supply.
- Government discourages entrepreneurs by taxing the risk-takers and redistributing the wealth to unnecessary civil employees, regulators, and unemployed "slackers." Therefore government is the impediment to progress. Steadily increasing population-demand and physical limits of supply (or the laws of physics) are totally ignored.

The Obama alternatives are increased fuel efficiency, "guarded" (environmentally sound) exploitation of remaining fossil fuels, and a transition to lower energy, renewable sources. A more recent White House initiative is: "How We Shift America Off Oil" (whitehouse.gov.energy).

The "Energy Security Trust" proposed spending ten billion dollars in the next ten years for lighter natural gas-vehicle fuel tanks, advanced batteries, cleaner biofuels, and hydrogen fuel cells. What is meant by "cleaner biofuels"? I thought the "Hydrogen Future" fairy tale had long-since been forgotten. All infer scientific "breakthroughs" that will solve our problems and negate the urgency for our nation to ration gasoline. Typically, neither left nor right political party dares to mention geological resource limits or the taboo subject of population.

Presidential election 2016

As we move into the hotly contested presidential election year, not a single candidate on either side seems to understand or dares discuss the overwhelming impact, facts, and future of the imminent end of the oil age. It would be amazing if a recipient of my 5th edition book could propel the urgency of our energy predicament far up the political ladder to someone (those) with a voice or position to make a difference. In 2005 Rep. Roscoe Bartlett (R Maryland) assumed that role of leadership. He held a Peak Oil conference in Frederick, Maryland where I presented. He personally gave a copy of my 2nd edition book to every congressperson. Now he is retired and a critical ten years have been lost.

The politics of population

Since the dawn of history, leaders have often been confronted by population growth that exceeded regional carrying capacity, especially when times were good. Increased numbers were often encouraged to swell the leader's kingdom and provide "boots on the ground" to ensure success, power, and defense. But in more recent times, several U.S. presidents openly promoted family planning after their terms in office, Eisenhower and Truman became honorary co-chairmen of the Planned Parenthood Federation. John F. Kennedy the first Catholic president, was instrumental in the allocation of millions of dollars to distribute contraceptives throughout the world. George H.W. Bush was nicknamed "Rubbers" during his term as a Republican Representative from Texas.

Now, going into 2016, as we continue on the trajectory of over-population, and already on the downhill side of per capita oil, any mention of the human-numbers problem is conspicuous by its absence. **It may well be that a democracy is not the best form of government in times (most often the case throughout history) of over-stressed carrying capacity. The plurality of numbers necessary to elect leaders are counterproductive to the hard leadership decisions and sacrifices necessary for survival.** Certainly, Ernest Shakleton, when his ship was trapped in the antarctic ice, did not defer to a voting majority or socialistic group-think. **He led and his men survived. We now desperately need this type of leadership to clearly explain to the public how dire the situation is and what drastic measures must be shared by everyone to reduce population, conserve oil, and nurture the earth's eco-systems.**

What next?

As argued many times in the preceding chapters, the only way to wind down and supercede our oil-based lifestyle is to **simultaneously** reduce oil consumption to ten percent of our current U.S. rate, **and begin the longer task of reversing population by reducing the average fertility rate to not more than one child per female**.

Obviously, no realistic person would expect this thinking to be promulgated or implemented, especially on a world-wide basis. However, it is idealistically possible that the U.S., which has significant oil reserves left and the most lifestyle to lose, might respect numerical facts, close its borders to the import-export of resources and people, and begin a modicum of survival in a post-oil world. All of this must happen in the next several decades to avoid the pending world-wide crash of modern civilization. **Half of the Americans alive today, including all new children from now on, will participate in this short epoch. This story must be broadcast far**

and wide, The remaining window of time is narrowing every day. Are we up to the task? We will soon know.

A prognosis

In my opinion, we will see the climax of high-energy civilization and our familiar comfortable lifestyles begin to unravel in the next five to ten years. This is the time frame for the most likely U.S. oil-depletion scenarios in Figure 2 to coinside with decreasing global net exports (GNE as explained in Chapter 4). This "Double whammy" will be impossible to ignore and begin to drastically change our lives. **Without some form of rationing, a return to high oil prices and a National Debt soaring past twenty trillion dollars will combine to destroy our economic stability.** Business as usual will end despite temporary remissions in the stock market and soaring dollar value with respect to other world currencies which are leading the way to economic collapse.

World-wide calamities will grow as terrorism reflects Mid-east unrest spreading from Syria, Egypt, Iraq, Iran, Saudi Arabia, Turkey, and into Africa, South America and Asia. **Throughout the world, more and more countries can no longer rely on cheap oil and/or oil exports to appease their growing generations of "oil babies" (children born throughout the world whose personal energy was rendered superfluous by ubiquitous, cheap oil).** Western nations will no longer be able to rely on oil-based-energy growth to continue the extrapolation of past prosperity. This, just as the Baby-boomers will be expecting their happy retirement years to be supported by entitlements, which in the past were paid for by ever-increasing growth in numbers from younger generations.

Conflict is certain at every level from the wealthy trying to preserve their assets and lifestyle, on down to increasing numbers of disenfranchised hungry protesters in the streets. Fuel for travel, farming, and heating will become prohibitively expensive for the growing majority who can only find menial work, or are dependent on shrinking economic safety nets. Continued lower-priced commodities will still be too costly for the number one consumer bloc: poor Americans, staggering under growing consumer debt. Civil and regional wars, exacerbated by religious or ethnic backgrounds will increase between nations that can no longer appease their own people and neighbors. **This is not a happy prognosis but is similar to predicting how many miles you can drive on a partial tank of gasoline before you walk.**

Bibliography

Listed below are most of the books in my library which bear directly on the interdependent subjects of energy, economics, peak oil and population. Most are available from amazon.com, some for as little as one cent plus shipping. Many of the titles use the words limits or collapse. Some are decades out of print. Obviously, I can't have read them all completely. I'm just a messenger. Time has run out. Is anybody listening?

Aleklett, K (2012) *Peeking at Peak Oil*

Astyk, S. (2008) *Depletion and Abundance, Life On the New Home Front.*

Baker, C. (2009) *Sacred Demise, Walking the Spiritual Path of Industrial Civilization's Collapse.*

Bartlett, A. (2004) *The Essential Exponential, For the Future of Our Planet.*

Bligh, J. (2004) *The Fatal Inheritance.*

Brown, et al. (1999) *Beyond Malthus, Nineteen Dimensions of the Population Challenge.*

Brown, L. (2008) *Plan B 3.0, Mobilizing To Save Civilization.*

Brown, L.(2011) *World on Edge, How to Prevent Environmental and Economic Collapse.*

Brown, L. (2012) *Full Planet, Empty Plates, The New Geopolitics of Energy Security.*

Bardi, U. (2011) *The Limits To Growth Revisited.*

Beck, L. (2009) *V2G-101.*

Berry, W. (1977) *The Unsettling of America, Culture and Agriculture.*

Borgstrom, G. (1965) *The Hungry Planet, the Modern World at the Edge of Famine.*

Boughey, A. (1976) *Strategy for Survival, An Exploration of the Limits to Further Population and Industrial Growth.*

Butler, (2012) *Energy: Overdevelopment and the Delusion of Endless Growth*

Campbell, C. (1997) *The Coming Oil Crisis.*

Campbell, C. (2003) *The Essence of Oil & Gas Depletion.*

Catton, W. (1982) *Overshoot, The Ecological Basis of Revolutionary Change.*

Catton, W. (2009) *Bottleneck, Humanity's Impending Collapse.*

Carroll, J. (1997) *The Greening of Faith, God, the Environment, and the Good Life.*

Carr-Saunders, A. (1922) *The Population Problem, A Study in Human Evolution.*

Cipolla, C. (1978) *The Economic History of World Population.*

Clark, W. (1974) *Energy for Survival, The Alternative to Extinction.*

Clark, W. (2005) *Petrodollar Warfare, Oil, Iraq, and the Future of the Dollar.*

Cohen, J. (1995) *How Many People Can the Earth Support?*

Cobb, K. (2010) *Prelude, A Novel About Secrets, Treachery, and the Arrival of Peak Oil.*

Cooke, R. (2007) *Detensive Nation, From Regulation to Leadership.*

Cox, S. (2013) *Any Way You Slice It, The Past Present and Future of Rationing.*

Cribb, J. (2010) *The Coming Famine, the Global Food Crisis and What We Can Do.*

Czech, B. (2000) *Fuel for a Runaway Train, Errant Economists, Shameful Spenders, and a Plan To Stop Them All.*

Daly, H. (1996) *Beyond Growth.*

Deffeyes, K. (2001) *Hubbert's Peak, The Impending World Oil Shortage.*

Deffeyes, K. (2005) *Beyond Oil, The View From Hubbert's Peak.*

Deffeyes, K. (2010) *When Oil Peaked.*

Diamond, J. (2005) *Collapse, How Societies Choose to Fail or Succeed.*

Daly, H. (1996) *Beyond Growth, The Economics of Sustainable Development.*

Dawkins, R. (2006) *The Selfish Gene, 30th Anniversary Edition.*

Dilworth, C. (2010) *Too Smart for Our Own Good, The Ecological Predicament of Humankind.*

Douthwaite, R. (1992) *The Growth Illusion, How Economic Growth has Enriched the Few, Impoverished the Many, and Endangered the Planet.*

Douthwaite, R. (2011) *Fleeing Vesuvius, Overcoming the Risks of Economic and Environmental Collapse.*

Ehrlich, P. (1971) *The Population Bomb.*

Ehrlich, P. And A. (2008) *The Dominant Animal, Human Evolution and the Environment.*

Fehling, G. (2015) *Dark Peak.* A new novel about life in a post-peak oil world.

Fletcher, S. (2011) *Bottled Lightning, Super Batteries, Electric Cars, and the New Lithium Economy.*

Friedrichs, J. (2013) *The Future is Not What it Used To Be.*

Gelbspan, R. (2004) *Boiling Point.*

Goodchild (2013) *Tumbling Tide: Population, Petroleum, and Systemic Collapse.*

Grant, L. (2000) *Too Many People, The Case for Reversing Growth.*

Grant, L.(2005) *The Collapsing Bubble, Growth and Fossil Energy.*

Greer, J. (2008) *The Long Descent.*

Greer, J. (2009) *The Ecotechnic Future, Envisioning a Post-Peak World.*

Greer J. (2013) *Not the Future We Ordered: The Psychology of Peak Oil and the Myth of Internal Progress*

Grover, J. (1991) *Beyond Oil, the Threat to Food and Fuel.*

Hardin. G. (1993) *Living Within Limits, Ecology, Economics, and Population Taboos.*

Hardin, G (1998) *The Ostrich Factor,Our Population Myopia.*

Hartmann, T. (1998) *The Last Hours of Ancient Sunlight.*

Hartmann, T. (2009) *Threshold: The Crisis of Western Culture.*

Hartmann, T. (2013) *The Crash of 2016.*

Heinberg, R. (2004) *Power Down, Options and Actions for a Post-Carbon World.*

Heinberg, R. (2005) *The Party's Over, Oil, War, and the Fate of Industrial Societies.*

Heinberg, R. (2007) *Peak Everything, Waking Up to the Century of Declines.*

Heinberg, R. (2006) *The Oil Depletion Protocol, a Plan to Avert Oil Wars, Terrorism, and Economic Collapse.*

Heinberg, R. (2009) *Searching For a Miracle, Net Energy Limits & the Fate of Industrial Society.*

Heinberg, R. (2010) *Post Carbon Reader, Managing the 21st Century's Crisis.*

Heinberg, R. (2011) *The End of Growth, Adapting to Our New Economic Reality.*

Heinberg, R. (2013) *Snake Oil, How Fracking's False Promise of Plenty Imperils our Future.*

Heinberg, R. (2015) *Afterburn, Society Beyond Fossil Fuels.*

Hirsch, R. et al. (2010) *The Impending World Energy Mess, What it Is and What It Means To You.*

Hopkins, R. (2008) *The Transition Handbook: From Oil Dependency to Local Resilience.*

Howe, J. (2006) *The End of Fossil Energy, and Last Chance for Survival.* (3rd Ed.)

Huber (2013) *Life Blood: Oil, Freedom, and the Forces of Capitalism.*

Huesemann (2011) *Technofix: Why Techonology Won't Save Use or the Environment.*

Jenkins, J. (2005) *The Humanure Handbook, a Guide for Composting Human Manure.*

Jenson, W. (1970) *Energy and the Economy of Nations.*

Karl, T. (1997) *The Paradox of Plenty, Oil Booms and Petro-States.*

Keen, S. (2011) *Debunking Economics, The Naked Emperor Dethroned.*

Klare, M. (2012) *The Race for What's Left, The Global Scramble for the World's Last Resources*

Kreith, F. (2014) *Sunrise Delayed, A Personal History of Solar Energy.*

Kunstler, J. (2005) *The Long Emergency, Surviving the Converging Catastrophes of the Twenty-First Century.*

Kuntsler, J. (2012) *Too Much Magic, Wishful Thinking, Technology, and the Fate of the Nation.*

Laslo, E. (2006) *Global Survival, the Challenges and its Implications for Thinking and Acting.*

Magdoff, F. (2010) *Agriculture and Food in Crisis, Conflict, Resistance, and Renewal.*

Malthus, T. (1798) *An Essay on the Principle of Population.*

Mass, P. (2009) *Crude World, The Violent Twilight of Oil.*

Martinson, C. (2011) *The Crash Course: The Unsustainable Future of the Economy, Energy, and Environment.*

Mesarovic, M. (1974) *Mankind at the Turning Point, the Second Report to the Club of Rome.*

McKibben, W. (1998) *Maybe One, a Personal and Environmental Argument for Single-Child Families.*

McPherson, G. (2013) *Going Dark.*

Meadows, D. (1972) *The Limits to Growth.*

Meadows, D. (2004) *Limits To Growth, The 30-Year Update.*

Moyo, D. (2012) *Winner Take All, China's Race for Resources and What it Means for the World.*

Nikiforuk, A. (2010) *Tar Sands, Dirty Oil and the Future of a Continent.*

Nikiforuk, A. (2012) *The Energy of Slaves, Oil and the New Servitude.*

Ophuls (2012) *Immoderate Greatness: Why Civilizations Fail.*

Orlov, D. (2008) *Reinventing Collapse: The Soviet Example and American Prospects.*

Orlov, D. (2013) *The Five Stages of Collapse: Survivor's Toolkit.*

Phillips (2006) *American Theocracy: The Perils and Politics of Radical Religion, Oil, and Borrowed Money in the 21st Century.*

Pimentel, D.(1996) *Food, Energy, and Society.*

Pfeiffer, D.(2003) *The End of the Oil Age.*

Ponting, C. (1991) *A Green History Of the World.*

Ravin, P & AAAS (2000) *AAAS Atlas of Population and Environment.*

Roberts, P. (2009) *The End of Food.*

Roberts, P. (2004) *The End of Oil, On the Edge of a Perilous New World.*

Romm, J. (2004) *The Hype About Hydrogen.*

Rothkrug, P. (1991) *Mending The Earth: A World For Our Grandchildren.*

Rubin, J. (2009) *Why Your World is About To Get a Whole Lot Smaller.*

Ruppert, M. (2009) *Collapse, The Crisis of Energy and Money in a Post Peak Oil World.*

Sabin, P. (2013) *The Bet, Paul Ehrlich, Julian Simon, and our Gamble Over Earth's Future.*

Sale, K. (1980) *Human Scale: Big Government, Big Business, Big Everything, How the Crises that Imperil America are the Inevitable Result of Giantism.*

Seidel, P. (1998) *Invisible Walls, Why We Ignore the Damage We Inflict on the Planet.*

Shepard, M. (2013) *Restoration Agriculture, Real-World Permaculture for Farmers.*

Simmons, M. (2005) *Twilight in the Desert, the Coming Saudi Oil Shock and the World Economy.*

Scheer, H. (1999) *The Solar Economy: Renewable Energy for a Sustainable Global Future.*

Smil, V. (1999) *Energies, an Illustrated Guide to the Biosphere and Civilization.*

Smil, V. (2005) *Energy At The Crossroads.*

Snyder-Meyerson (2015) *Peak Oil: Apocolyptic Environmentalism and Libertarian Political Culture.*

Stanton, W. (2003) *The Rapid Growth of Human Populations.*

Tainter, J. (1988) *The Collapse of Complex Societies.*

Tamminen, T. (2006) *Lives Per Gallon: The True Cost of Our Oil Addiction.*

Weeks, J. (2005) *Population, an Introduction to Concepts and Issues.*

Weisman, A. (2007) *The World Without Us.*

Weisman, A. (2013) *Countdown: Our Last, Best Hope for a Future on Earth.*

Wilkinson, R.(1973) *Poverty and Progress.*

Wilson, E. (2002) *The Future of Life.*

Yetiv, S. (2015) *Myths of the Oil Boom: American National Security and a Global Energy Market.*

Young, L. (1968) *Population in Perspective.*

Youngquist, W. (1997) *Geodestinies, the Inevitable Control of Earth Resources Over Nations and Individuals.*

And a couple that just don't get it:

Cole, H. (1973) *Models of Doom, a Critique of the Limits to Growth.*

Huber P. et al. (2005) *The Bottomless Well: The Twilight of Fuel, the Virtue of Waste, and Why We Will Never Run Out of Energy.*

Lacalle and Parrilla (2015) *The Energy World is Flat: Opportunities from the End of Peak Oil.*

LaRouche, L. (1983) *There Are No Limits To Growth.*

Zehner, O. (2012) *Green Illusions: The Dirty Secrets of Clean Energy and the Future of Environmentalism.*

APPENDIX A

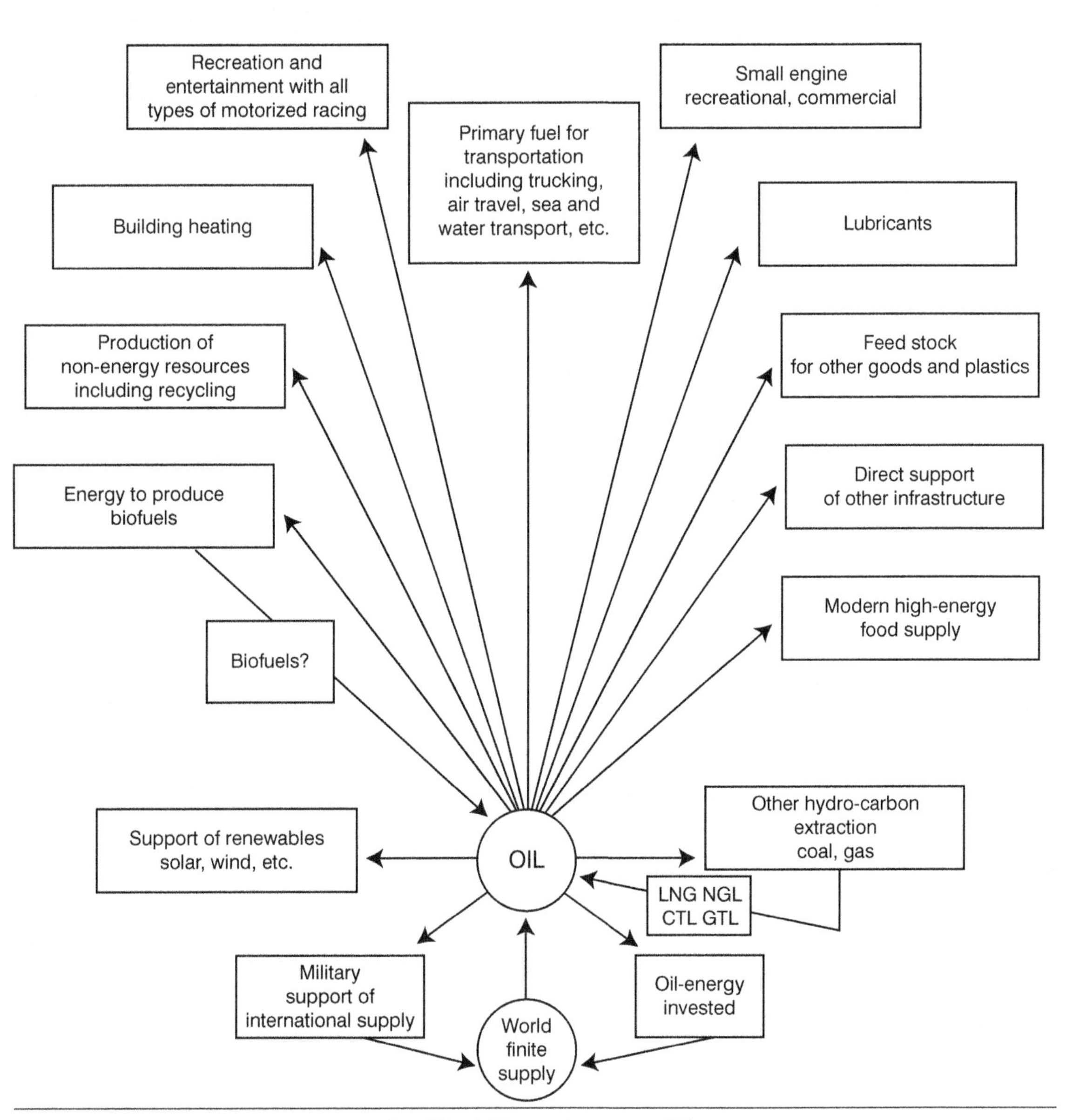

Our Total Dependency on Oil

APPENDIX B

Energy 101 … Energy, Work, Power

The following basic definitions, energy equivalences, and necessary explanations will be helpful as background for this book or other references:

ENERGY is the ability to do work.

WORK is the movement, of something from one place to another, or the equivalent amount of heat energy required to raise something from one temperature to a higher temperature.

POWER is how fast the work happens or the rate that the energy is being used.

Work (W) can be expressed as a force (F), like a push or a pull, occurring over a distance (D) and in the direction traveled. For instance, if it takes 10 pounds of push or pull to provide enough sliding force to move a sled, and the distance traveled is 50 feet, the force times the distance equals the work.

In equation form:

$$\text{Force} \times \text{Distance} = \text{Work}$$

With the numbers substituted in our sled example:

$$10 \text{ lbs.} \times 50 \text{ ft.} = 500 \text{ ft.lbs.}$$

Note, that if a 9 pound force was exerted against the sled and no movement occurred, no work would be done.

$$9 \text{ lbs.} \times 0 \text{ ft.} = 0 \text{ ft.lbs.}$$

You could lean (or prop a stick in your place) against the sled all day, the sled would not move and no work would be done.

Work requires both force and distance traveled. Energy, which can be in different forms, is the ability to do the work. Power is how fast the work is done.

There are many ways that the sled could be moved, such as with a snowmobile, a dog team, cross-country skier, a winch with an electric motor, a sail, and many more, but in each case, the sliding work done and energy used is the same, 500 foot-pounds, as long as the sum of the forces resisting movement totals 10 pounds.

If there were wheels on the sled so it rolled and required only 5 pounds of force, then the work done would be 250 foot-pounds (5 lbs. × 50 ft. = 250 ft.lbs.) The wheel was a great invention to reduce the work of transportation and lessen the required energy.

Energy exists in many different forms, but in each case suggested above, the energy required, or the ability to do the work of overcoming 10 pounds of drag, is still 500 foot-pounds. It doesn't matter if the force came from a gasoline engine, an animal, a human, or the wind. Energy is that elusive something able to do the work. The units for energy and work are the same.

At this point, we will leave energy for a minute and discuss power so that we are absolutely clear about the three terms: energy, work, and power.

It doesn't take a great leap of intuition to appreciate that moving the sled in 50 seconds is a much different task than moving it the same 50 feet in five seconds. A dog or person could easily handle the 50-second task, but it would take a snowmobile or a good gust of wind with a big sail to move it the same distance in five seconds. Power is the time rate of doing the work. In equation form:

$$\text{Power} = \text{Force} \times \text{Distance/Time}$$

Our original slow sled example becomes:

$$P = 10 \text{ lbs.} \times 50 \text{ ft./50 sec.} = 10 \text{ ft.lbs./sec.}$$

In the faster example such as with wind or snowmobile:

$$P = 10 \text{ lbs.} = 50 \text{ ft./5 sec.} = 100 \text{ ft.lbs./sec.}$$

This is ten times as much power as the slower example.

The work in each case is the same, but the power is ten times greater because the time (T) to do the work changed from a task we could easily do ourselves slowly to a ten times quicker task requiring a machine stronger than most humans. Remember, the work done and the energy used while moving between two points, no matter what the time, is the same assuming the sliding friction remains the same in each case.

If we're in a hurry, we need more power. A stronger machine or an athlete can do the same work as the weaker but in less time. Power is a measure of the effort within a specific time period.

A common English unit for power is horsepower (Hp). By definition one horsepower equals 550 foot-pounds of work done in one second. To introduce horsepower to our above slower sled problem:

10 ft.lbs./sec. divided by 550 ft.lbs. per second (per Hp) = 0.018 Hp

In the faster case, the horsepower is:

100 ft.lbs./sec. divided by 550 ft.lbs. per second (per Hp) = 0.18 Hp

---

In real life, as speed increases, the air resistance will start to be a significant drag factor in addition to the sliding friction. Air resistance is a squared factor. Even though it might be negligible at slow speeds it increases four-fold each time the speed doubles.

In the second case, when moving the sled ten times as fast, the air resistance would be 10^2 or 100 times greater and in theory start to add to the sliding friction depending on the speed, shape, and the frontal area of the sled. Together, air resistance plus sliding friction make up a total drag force (F) and the amount of work (F × D) required to move the sled. This is why a streamlined shape becomes so important at high speed. Air resistance explains why moving things in a hurry is much more energy intensive. However, in our sled examples, the faster case of a 50-foot distance in 5 seconds is 10 ft./sec., or only about 7 mph. This is still slow enough that air drag is insignificant compared to the sliding friction so it can be ignored.

The energy needed to overcome air resistance starts to become important above 20 mph. At higher speeds, movement through air becomes the very dominate drag force resisting most of the work (energy).

In order to complete this short review of energy, work, and power, we need to clarify different forms of energy as well as to gain a quantitative understanding of the amount of work each can do.

MECHANICAL ENERGY

To review traditional mechanical terms described above:

Work = distance times force = foot pounds

Energy is equivalent to work = foot pounds

Power = work divided by time = foot pounds per second

THERMAL (the subject of heat) ENERGY

To raise a mass of material a certain temperature also requires energy. Therefore, work, the energy required, and the temperature increase of a substance are equivalent. I will try to clarify this concept with a few definitions.

The familiar term BTU (British Thermal Unit) is the amount of energy (or work) required to raise one pound of water one-degree Fahrenheit. To heat one gallon (about 8 pounds) of water fifty degrees would require:

$$8 \text{ lbs.} \times 50°\text{f} = 400 \text{ BTU}$$

The BTU is a term of energy (or work) just like a foot-pound. One BTU is directly equivalent to 778 foot-pounds.

In metric terms, one **calorie** is the amount of energy (or work) required to raise one gram of water one degree centigrade. To heat the same gallon of water as above, which weighs 8 pounds (3632 grams), 50°f (27.8°c) would require:

$$3632 \text{ grams} \times 27.8°\text{c} = 100{,}970 \text{ calories}$$

Obviously one calorie is not very big, so scientists sometime use the kilocalorie (1000 calories) especially as a measure of energy value of food. When this is done, a capital C may be used (not to be confused with temperature in degrees centigrade, °c).

Thermal energy may also be the output of a chemical reaction. This occurs when the complex-carbon molecules of fuel combine with oxygen to make heat (thermal energy) and simpler, lower energy, carbon molecules like carbon dioxide (CO_2) or carbon monoxide (CO). This process is called burning or combustion.

The minimum food energy required by a healthy adult in one day is about 2,000 kilocalories. Humans are very complex machines requiring large quantities of sophisticated fuels. All basic carbohydrate-fuel sources provide about 4 kilocalories/gram of energy. The average requirement of 2,000 kilocalories per day per person equals 500 grams (just over one pound) of carbohydrates per day or about 400 pounds per year. Proteins have about the same caloric-energy value and may be a substitute for carbohydrate energy. Fats have two to three times the energy value of carbohydrates.

Further reference about kilocalorie counting for food is available on the Nutrient Data Laboratory website (www.nal.usda.gov).

We can now see why food production, as the fuel source for our bodies, is the primary energy concern for any society. It takes considerable energy (fossil fuel, animal power, and/or human power) to produce 400 pounds of food per year per person. With today's energy intensive modern agriculture, 10 to 20 kilocalories are used to deliver one kilocalorie of food to the dinner table. Without the fossil fuels or some other alternative source of more concentrated energy, we will have to go back to subsistence lower-yield agriculture using human and animal power to provide our food.

Now that we have enough basics of scientific terms and quantification of energy, we can better appreciate the astonishing energy content of fossil fuels and reject inaccurate or misleading information. Consider the following facts straight from any technical source book, and compare the different concentrated-energy sources with the feeble output of human labor.

One gallon of oil, kerosene, diesel fuel, gasoline, fat, etc. (they're all about the same) has the concentrated combustion energy of approximately 150,000 BTU. This is a tremendous amount of stored energy and represents the ability (before efficiency losses) to do over one hundred and sixteen million foot-pounds (150,000 BTU per gallon × 778 ft.lbs. per BTU = 116,700,000 ft.lbs. per gallon) of work. We can go down to the corner gas station and buy this amount of fuel (energy) for about $1.50. Working at a typical continuous maximum of 256 BTU per hour of power (75 watts), hour after hour, a strong healthy adult would have to work 600 hours to equal this amount of energy. No wonder we live in a age of comparatively "free" energy. Even if the fuel were converted into work at a 25 percent efficiency rate (typical for fossil-fuel energy), it would still take 150 hours of steady manual labor to equal the energy in a gallon of fossil fuel.

At $1.50/gal. for gasoline, equivalent human labor to do the same work is worth about one cent per hour. Soon to be $3.00 per gallon gasoline would still make the equivalent human labor worth only 2 cents per hour.

A similar analysis reveals that a pound of coal with 10,000 BTU is equal to about 10 hours of manual labor and a pound of dry wood with 5,000 BTU per pound contains energy equal to five hours of constant manual labor.

ELECTRICAL ENERGY

Unfortunately, for the average person trying to understand the subject of energy, the electrical community has still another set of terms for energy, work, and power. The electrical engineer starts with the basic concepts:

Volts (the force, or push, like pounds in mechanical terms) times the current (in units of amperes or the quantity of electrons flowing at the speed of light) equals watts (power).

$$\text{Volts} \times \text{Amps} = \text{Watts}$$

Since we have force, distance, and time (understood as speed of light), we have power. The watt (or kilowatts as thousands of watts) is the fundamental unit of electrical power in both English and metric systems. The equivalence of electrical and mechanical power is defined as:

$$746 \text{ watts } (0.746 \text{ kw}) = \text{one horsepower}$$

One kilowatt and three-fourths of a horsepower are similar in magnitude. In Europe, cars are rated in kilowatts instead of horsepower. Since a kilowatt is a unit of power, it must be multiplied by time to get back to the simpler concepts of work and energy like foot-pounds. The common product of kilowatts times hours is a measure of how much we pay for electrical energy coming into our homes from the utility grid. One kilowatt (power) times one hour = one kilowatt hour:

$$1 \text{ kw} \times 1 \text{ hr.} = 1 \text{ kwh}$$

In your monthly electrical bill you will see that electrical energy (not the power) costs about 15 cents per kilowatt-hour. A typical household may use about 700 kwh per month for about $105.00. This is an incredible bargain unique to our modern, low-cost energy, industrialized civilization.

A strong human working a complete 40 hour week can only produce 75 watts of power times 40 hours, which is equal to 3000 watt hours (3 kwh) each week. This is equivalent to the thermal energy required to heat one hot shower. At electrical costs of 15 cents per kwh, a week of hard human work would be worth about 45 cents, about one cent per hour.

The above technical background is all we need to understand the remainder of this book and the magnitude of the situation we've gotten into in just the last 150 years.

The following is offered as a brief review of the history of energy:

5000 B.C.—About this time, human beings came out of the woods or savannahs as hunter/gatherers in delicate balance with nature and started to grow a little excess food energy in the form of grains. This slight surplus of energy beyond the minimum required for survival allowed the beginning of civilization. In addition, beasts of burden were domesticated to add to the amount of work that humans alone could do as well as provide concentrated food-energy as meat and milk. The agricultural revolution allowed humans to stop moving about, build villages, and multiply in population one hundred fold from about 10 million to almost a billion by the start of the Industrial Age. This early age of grain-energy reached a peak when the expansion and military might of Rome extended to the limits that grain production could support. The use of slave labor in all "civilized" countries provided incremental energy to do additional work. This practice with its detestable human rights issue is common in biblical references and continued right into the fossil-fuel era beginning 150 years ago and even to the present in some parts of the world. The U.S. was founded on vast resources and slave labor.

500 A.D.—Non-fossil-fuel machines like water wheels and windmills were built to further energy availability and give another small boost to civilization and leisure time.

1500—Civilization continued to grow slowly until wood, which was the major source of fuel for energy needs, began to be seriously depleted. Fortunately, the first use of fossil fuels in the forms of coal and peat satisfied ever increasing energy demands and kept the quality of life gradually improving. During this period the whale population was almost decimated in a few decades for the oil to be used in lighting more extravagant homes.

1775—James Watt (1736–1819) invented the steam engine and ushered in the Industrial Age by improving on the piston pump used for removing water from coal mines. This wondrous machine allowed the conversion of concentrated, previously stored energy, such as wood and coal, into work and power far in excess of what man, beast, windmill, or waterwheel could do.

1859—About the time of the Civil War when coal driven locomotives and steamships were already well established, Colonel Edwin Drake struck plentiful crude oil in Pennsylvania. This drilling technology was rapidly followed by further discoveries in the U.S. and then eventually worldwide. Plentiful liquid crude oil eventually supplanted coal as the number one fossil-fuel energy source because of its ease of procurement, transportation, and utilization in machines which could convert its huge energy content into useful work.

1900—Beginning about a century ago, inventors found many new ways to harness the concentrated energy of free flowing oil and cleaner derivatives like kerosene and gasoline. All types of machines evolved to power industry, agriculture, and especially transportation for military, personal, and commercial use on land, sea, and in the air. There are centenarians alive today that span this entire era. It is no small coincidence that the 100-year anniversary of the airplane exactly coincides with this hydrocarbon energy epoch. Oil became an absolutely essential military force in WWI as it provided submarines, airplanes, tanks, and ships that could be refueled at sea.

The tremendous power requirements to move large masses quickly over long distances can only be provided by fossil fuels. The only exceptions are concentrated biofuels and liquid hydrogen, both requiring even larger energy inputs for their formation than they return for useful work.

1950—The use of fossil fuels including natural gas to make hydrogen culminated in more recent years as rocket fuel. This is the only way to provide the awesome power required to propel objects outward against the pull of the very strong force of earth's gravitational field. (The use of hydrogen as a concentrated fuel in lieu of fossil fuel will be discussed in more detail in other parts of this book.)

❧

At this point I will summarize the units and equivalents used for energy, work, and power into one table sufficient for understanding the remainder of this book and other references. Wherever possible, mechanical energy, electrical energy, and work will be quantified as kilowatt hours, and power will be in kilowatts. Thermal energy will be referred to in terms of BTU's or equivalent billion barrels of oil (EBBO), with each barrel containing 42 gallons or 6,300,000 BTU's worth of energy. Keep in mind that the total thermal energy of a fossil fuel cannot be converted directly to useful electrical energy, mechanical energy, or work without an energy loss. A 25% efficiency factor for internal combustion engines and 35% for electric power plants will be used where appropriate to calculate the equivalent fossil-fuel energy. In other words, usable, secondary electrical or mechanical output is divided by the efficiency to find the required primary energy. For instance a ten-kilowatt photovoltaic system would equal a fossil-fuel, power-plant input of 10/0.35 or 28.57 kilowatts equal to 97,486 BTU/hr.

SUMMARY of EQUIVALENTS and UNITS

1,000 = 10^3 = thousand (or kilo, K)

1,000,000 = 10^6 = million (or mega, M)

1,000,000,000 = 10^9 = billion (or giga, G)

1,000,000,000,000 = 10^{12} = trillion (or tera, T)

1,000,000,000,000,000 = 10^{15} = quadrillion (or peta, P)

Energy (or work equivalent)

1 kwh = 3412 BTU

1 barrel of oil = 1846 kwh = 6.3×10^6 (million) BTU

1 kilocalorie = 3.968 BTU

1 BTU = 778 foot pounds

1 foot pound = 1.356 Joules (Newton meters)

1 BTU = 0.252 kilocalories

1 kwh = 860 kilocalories

1 kwh = 2.65×10^6 (million) foot pounds

1 gallon hydrocarbon fuel = 150,000 BTU

1 gallon biodiesel = 121,000 BTU

1 pound high quality coal = 10,000 BTU

1 pound dry wood = 5,000 BTU (wet wood may have zero energy value)

1 cubic meter (36 cubic feet) natural gas = 36,000 BTU

1 cubic foot natural gas = 1,000 BTU

1 trillion (10^{12}) cubic meters natural gas = 36 quadrillion (10^{15}) BTU

1 trillion (10^{12}) cubic meters natural gas = 5.72 EBBO (equivalent billion barrels of oil)

1 quadrillion (10^{15}) BTU = 0.159 EBBO

Power

1 horsepower (Hp) = 550 foot pounds/second (ft.lbs./sec.)

1 kw = 1.34 Hp

1 Hp = 746 watts (0.746 kw)

1 kw = 0.95 BTU/sec.

1 Hp = 0.71 BTU/sec.

1 watt = 1 Joule/sec. = 1 Newton meter/sec. (metric terms used for power)

CPSIA information can be obtained
at www.ICGtesting.com
Printed in the USA
BVOW10s0252290317
479456BV00003B/9/P

9 780996 205450